Handbook for the Field Assessment of Land Degradation

Michael Stocking and Niamh Murnaghan

Earthscan Publications Ltd, London • Sterling, VA

First published in the UK and USA in 2001 by
Earthscan Publications Ltd

ISBN: 1 85383 831 4

Typesetting by PCS Mapping & DTP, Newcastle upon Tyne
Printed and bound by Thanet Press, Margate, Kent
Cover design by Danny Gillespie

For a full list of publications please contact:

Earthscan Publications Ltd
120 Pentonville Road, London, N1 9JN, UK
Tel: +44 (0)20 7278 0433
Fax: +44 (0)20 7278 1142
Email: earthinfo@earthscan.co.uk
http://www.earthscan.co.uk

22883 Quicksilver Drive, Sterling, VA 20166–2012, USA

A catalogue record for this book is available from the British Library

Library of Congress Cataloging-in-Publication Data applied for

Earthscan is an editorially independent subsidiary of Kogan Page Ltd and publishes in association
with WWF-UK and the International Institute for Environment and Development

This book is printed on elemental chlorine-free paper

Table of Contents

List of Figures

List of Tables

List of Boxes

List of Plates

(see colour plate section)

Preface

This handbook is designed to provide assistance to those interested in collecting measurements and assessments of land degradation rapidly in the field. It has a particular emphasis on the effects important to land users and a special focus on dialogue with farmers who can both advise on what is important to them and also give the field assessor a continuous monitoring capability which would otherwise be missed in occasional field visits. Primary consideration is given to small-scale rainfed agriculture in the tropics because this covers the majority of situations and the largest number of rural people. While large-scale commercial agriculture is not specifically mentioned, and rangeland and wetlands only briefly so, the principles that apply throughout this handbook will be of assistance.

The need, expressed to us many times by field workers, for a readily accessible and practical guide to field measurement of land degradation inspired the writing of this book. Standard scientific techniques have usually involved bounded field plots and measurements of soil loss and runoff into collecting tanks. But these are cumbersome methods, yielding only limited information even after several years of monitoring. The artificiality of the experimental devices also renders many of the results difficult to interpret in a way meaningful to real field conditions. So, when we have undertaken fieldwork with our collaborators, most of whom are from (and work in) developing countries, we have been on the alert for simple, direct and useful measures of the dynamics of the processes leading to land degradation. We have found that the more we have looked, the more we have uncovered evidence in the field that has been unseen in the past. The evidence may only amount to small accumulations of soil, or thin layers of residual stones on the surface, both easily overlooked. However, these are 'real' pieces of evidence occurring in actual fields being used by farmers; they represent the outcomes of processes usually instigated by land use practices; and they are often the very processes and evidence that farmers themselves also see. So, we feel, such evidence has enormous value – a value that is enhanced by the fact that many measurements can be accomplished much more rapidly than by the standard techniques. If this value can also be linked to engaging with the land user in a dialogue on how the problems may be reversed, then we have a far better basis for positive change in the landscape. Although Rapid Rural Appraisal (RRA) and Participatory Rural Appraisal (PRA) have tended to be dominated by social or economic enquiry, we believe that change in natural resource quality is also amenable to the benefits of RRA and PRA approaches. The use of techniques that bridge the natural and social sciences is an aspect of fundamental importance to the field assessor that this handbook promotes.

Land degradation is a topic that is regaining prominence. Because of its potential threat to land resources and to the viability of human societies, land degradation has been the subject of alarming statistics. For example, the Global Assessment of Land Degradation (GLASOD) project calculates that 22.5 per cent of all productive land has been degraded since 1945, and that the situation is becoming rapidly worse. Yet, at the same time, few people have a clear idea of what land degradation is and even fewer could suggest ways in which it can be practically assessed in the field.

The confusion is unsurprising. Land degradation has tended to become caught up with other debates on environmental change. Degradation is, however, a biophysical process well known to farmers and other land users. Routinely, they describe how soils are getting thinner and 'worn out' and how yields are declining. As degradation progresses, farmers' efforts to secure a living become increasingly precarious and uneconomic. This book will focus exclusively at this level – on assessing degradation as a process affecting the activities of the farm household, rather than attempting global, national, regional or provincial assessments. Efforts to extrapolate to larger areas of land than the field or farm are fraught with inaccuracies and dubious assumptions, which we shall leave to others. Our focus will be through the eyes of farmers (see Chapter 1), addressing issues that concern land users as of primary importance (see Chapter 3). In Chapter 2 we shall carefully distinguish between land degradation, aspects of it such as soil degradation, and some of the biophysical processes that lead to land degradation. Inevitably, indicators will have to be used, and many of these will be derived from degradation processes such as soil loss (see Chapter 4) or degradation outcomes such as the effects on production (see Chapter 5). Assessments of land degradation are not, by themselves, very useful. Therefore, we show how the simultaneous collection of several indicators can lead to a much better realization of the relevance to land users (see Chapter 6), showing the consequences (see Chapter 7) and giving leads into the design of appropriate techniques of conservation (see Chapter 8). It is not, however, our objective to present conservation options; many technologies exist and handbooks on them abound.

Acknowledgements

We are grateful to two projects that have given us the opportunity to bring our experiences together into this handbook of field techniques. First, the People, Land Management and Environmental Change (PLEC) project, funded by the Global Environment Facility 1998–2002, implemented by the United Nations Environment Programme (UNEP) and executed by the United Nations University (UNU) in Tokyo, has a global network of demonstration sites. Farmers demonstrate 'good practice' on these sites in managing and conserving biological diversity. Part of this management relates to maintaining soil quality, and preventing land degradation. Hence, PLEC collaborators (now numbering over 200 in 12 developing countries) have been making field assessments of land degradation to support their monitoring of examples of good practice. Michael Stocking is the Associate Scientific Coordinator of PLEC and our two advisers have also been consultants for the project; Geoff Humphreys has a particular role in undertaking land degradation assessment. UNEP and UNU have requested additional support and guidance for these field activities, and this handbook is intended to provide them.

Second, the UK Department for International Development (DFID) funded a research project in 1996–1999 in Sri Lanka under its Natural Resources Systems Programme, entitled Economic and Biophysical Assessment of Soil Erosion and Conservation (R6525), which developed a number of the techniques described in this book. Michael Stocking was the Principal Investigator, and the project involved many Sri Lankan hill farmers showing how they perceived soil erosion and how land degradation was perceived by them. Rebecca Clark was the ODG Research Associate for this project, and we are grateful to her for working with many of the techniques in this handbook in the field and for helping to develop a clear farmer-perspective. DFID also commissioned the project to undertake a training course on soil erosion assessment in Bolivia in 1998, attended by some 30 local professionals, in which many of the techniques were tested. In Sri Lanka and Bolivia our local collaborators in the field became excited as they saw more and more evidence of degradation in field drains, boundary walls, under stones and in the middle of fields. Even an experienced soil surveyor said that he was seeing things he had not noticed before in 30 years of fieldwork. We want to try to transmit that enthusiasm to others through this book. We are extremely grateful to both UNU/UNEP and DFID and to our many collaborators. This handbook is officially an output from both projects. We would also like to thank the Nova Development Corporation for their kind permission to reproduce the clip art images that appear in the book.

However, without funding support from UNEP through trust funds from the Government of Norway, we would have been unable to collate the many experiences, photographs and measurement techniques that formed the basis of the first draft of the book. Timo Maukonen at

UNEP has been most supportive of this project, and we thank him sincerely. His enthusiastic comments on an early draft gave us great encouragement. Still, however, the draft needed to be fully tested and evaluated. The Natural Resources Systems Programme (NRSP) of DFID stepped in again and funded a full field workshop in Uganda, from which the reader will see that a number of examples has been drawn for this final version of the book. NRSP has also kindly subsidized 200 copies of the book for free distribution to developing countries, and a video of the techniques as witnessed in the field at Bushwere in southwestern Uganda. Our Ugandan colleagues' input and enthusiasm was unstinting, even though many are in senior and responsible posts in education and government. They wanted readers to know that they are very willing to be involved in future field assessments of land degradation, and so we are happy to mention them all by name and warmly thank them for their essential contribution to the final work. From Makerere University we would like to thank Joy K Tumuhairwe, Professor Julius Zake, Christopher Gumisiriza, Charles Nkwiine, Godfrey Taulya, Pamela Busingye and Giregon Olupot (all in Soil Science), John Kawongolo and Peter Mulamba (in Agricultural Engineering) and Geoffrey Lamtoo (in Environment and Natural Resources). From the National Agricultural Research Organisation we would like to thank Dr Henry Ssali. From the National Environment Management Authority we are grateful to Dr Festus Bagoora. From the Ministry of Agriculture, Animal Industry and Fisheries we owe thanks to Emmanuel Mpiirwe, Aloysius Karugaba, Sandra Mwebazi and Dan Miiro. Finally, from Mbarara University of Science and Technology we would like to thank Pamela Mbabazi, Charles Mucunguzi and Dominic Byarugaba.

In addition, we must mention our two advisers on the project: Anna Tengberg at UNEP has worked with us on land degradation issues for several years and has given us valuable advice; Geoff Humphreys of Macquarie University has provided training materials from his work for PLEC as well as additional material from his own work in Australia. We thank them both for their interest and dedication.

We also wish to thank those people who kindly reviewed the draft handbook and gave us valuable suggestions, which we have endeavoured to incorporate in the final text. Providing admirable critical comment (in date order of receipt) have been: Christine Okali (ODG/UEA and HTS Development, UK), Francis Shaxson (ex-FAO and DFID, UK), Frits Penning de Vries (IBSRAM, Thailand), Malcolm Douglas (consultant, UK), Harold Brookfield (ANU and PLEC Coordinator, Australia), Libor Jansky (UNU, Japan), Mario Pinedo Panduro (Instituto de Investigaciones de la Amazonia Peruana, Peru), Karl Herweg (Centre for Development and Environment, Switzerland), Will Critchley (Free University, The Netherlands), John McDonagh (ODG/UEA, UK), Samran Sombatpanit (retired from Department of Land Development, Thailand), Tej Partap (ICIMOD, Nepal), Ibrahima Boiro (Université de Conakry, Guinée Republique) and Dominique Lantieri, (FAO, Italy). While having given freely of their time and intellect, our colleagues are not to be blamed for failings in the final product. We happily invite additional observations from those who have tried the techniques in this handbook in the field – hopefully one day we shall revise the book to make it more practical and useful for all practitioners dealing with the problems of land degradation and its impact on human society in a field setting.

Michael Stocking
Niamh Murnaghan
Norwich, April 2001

List of Acronyms and Abbreviations

ANU	Australia National University
DFID	Department for International Development (UK)
FAO	Food and Agriculture Organisation
HTS	HTS Development Limited
IBSRAM	International Board for Soil Research and Management
ICIMOD	International Centre for Integrated Mountain Development
ITK	indigenous technical knowledge
NPV	net present value
NRSP	Natural Resources Systems Programme (DFID)
ODG	Overseas Development Group (University of East Anglia)
PLEC	People, Land Management and Environmental Change
PRA	Participatory Rural Appraisal
RRA	Rapid Rural Appraisal
UEA	University of East Anglia (UK)
UK	United Kingdom
UNEP	United Nations Environment Programme
UNU	United Nations University

Gaining a Farmer-perspective on Land Degradation

Introduction

Land degradation manifests itself in many ways. Vegetation, which may provide fuel and fodder, becomes increasingly scarce. Water courses dry up. Thorny weeds predominate in once-rich pastures. Footpaths disappear into gullies. Soils become thin and stony. All of these manifestations have potentially severe impacts for land users and for people who rely for their living on the products of a healthy landscape. These perceptions of land degradation are the typical views of observers concerned about a deteriorating environment and the declining livelihoods of land users.

Local people, however, may see the degradation in entirely different ways. For example, a woman increasingly engaged in collecting firewood and fetching water will worry about the scarcity of these natural resources and the burden of having to travel long distances to gain them. A herder of livestock in the same village will have concerns in searching for elusive dry-season pastures. A farmer will notice damage to the plough caused by the increasing litter of stones on the soil. Another person may not perceive the degradation to be a problem as she adapts her home-garden practices to the needs of the family. So, there are different perspectives within local social contexts, which need to be reflected in any field-level assessment of land degradation.

In making assessments of land degradation, the assessor is faced not only with the differing perspectives of land users but also his or her own perspective. Land users prioritize various aspects of degradation quite differently from local professionals or expatriate experts. They 'see' things which concern them and affect their way of life. This contrast in perspectives between those of scientifically trained professionals and those of local people is difficult to tackle because it involves ourselves and our own prejudices. Science teaches us that we are right: the setting of hypotheses, the experimental testing of alternatives and the analysis of process – all are intended to verify what is actually happening and to prove cause and effect. However, the products of science have not always been 'right'. Developing countries are littered with technologies that have been promoted and have failed. Technical recommendations that have been thoroughly researched are frequently rejected by local people.

Nowhere is the failure of technology more evident than in soil and water conservation – the main antidote to soil degradation. Since the 1940s, mechanical techniques of soil conservation such as broad-based terraces have come and gone; biological techniques such as strip cropping were popular in the US in the 1960s and are now hardly ever seen. Even conservation tillage, hugely effective in reducing soil loss by keeping residues on the surface, has only been taken up by a minority of farmers in places where it has been strongly promoted. There is clearly a mismatch between the perspectives of the scientists, technology developers and local professionals, and the views of land users who are expected to implement the recommendations.

This book addresses the mismatch. It asks whose views are more important: the land user, who suffers the consequences of land degradation and who may benefit (or not) from conservation; or the professional, whose livelihood does not depend on the land? This handbook promotes the likely views of land users, not just as a different perspective, but as a set of views of land degradation that is much more relevant to the design and promotion of acceptable technologies. Of course, it is difficult to 'put oneself in another's shoes' and to subjugate one's own strongly held views, but it is worth the effort, if only to get a better assessment of which interventions will bring lasting benefit.

To explore the differences in perspective further, consider erosion-induced loss in soil productivity. This biophysical process, whereby soil erosion reduces the quality of the soil and hence its ability to produce vegetation, is one of the key issues in debates on food security. If degradation is reducing current and future yields, the argument goes, future populations will not be able to feed themselves. If future yields decline, farming will become increasingly difficult (see Plate 1). Erosion-induced loss in soil productivity may occur through a variety of processes, described in partially scientific terms – ie the professional perspective:

- Loss of nutrients and organic matter in eroded sediments reduces the total stock of nutrients in the remaining soil that will be available to future crops

- Reduction in plant-available water capacity, through the selective depletion of organic matter and clays by erosion, increases the chances of drought stress in future crops
- Increase in bulk density, surface crusting and other physical effects of soil degradation prevent seed germination and disrupt early plant development
- Reduced depth of topsoil and exhumation of subsoil by long-term soil erosion decrease the available soil volume for plant roots
- Increasing acidity through selective removal of calcium cations on the exchange complex affects nutrient availability, encourages P-fixation and induces free aluminium, causing severe toxic effects
- Reduction in micro-faunal and micro-floral populations affects beneficial processes, such as nitrification
- Because of poorer soil properties, loss of seeds and fertilizers, poor germination and in-field variability and other direct process effects of degradation, farming operations become more difficult and less economic.

These processes present a complicated, interactive and cumulative picture of how land degradation may translate to an actual decline in farm production. Only some of these individual processes may be recognized by farmers. Reduced soil depth and poor seed germination are often cited, but rarely do farmers relate them directly to erosion. Other processes, such as aluminium toxicity causing massive crop failure, usually go unrecognized. Since this occurs mainly in the humid tropics where farmers have shifted their plots after only a few years of cultivation, the failures will be seen as nothing more than the normal course of events. Farmers have described their degraded soils as 'worn out', 'no good' or 'weed-infested'.

A farmer's perspective, therefore, will usually be different from, and the ascribing of cause and effect quite unrelated to, the scientific explanation. The classic example of this is the explanation of soil formation by the Burungee of Tanzania, as discovered by the anthropologist Wilhelm Östberg. The Burungee see stones on the surface of the soil. To them it is evidence that 'the land is coming up' and that soil formation is active. To the scientist, stones are the residual left after erosion, and are clear evidence of the very opposite of soil formation (see Plate 2).

Differences in thinking and explanation are not always as stark, but can be every bit as powerful. Take the response of a Tanzanian farmer to a question posed by Per Assmo about rills and erosion: 'These small rills in the field are not erosion. That is only water running through the field. The rills disappear during field prepara-tions' (1999, p148, see Appendix IV). Such descriptions give fascinating insights into farmers' explanations and priorities. Soil productivity is articulated principally through the *consequences* of its change, what farmers see going on in their fields, and the effect that this has on farming practice and production.

This handbook will adopt the evidence of land degradation in the field through what farmers have said they see, the effects that they have described, and how their farming practices have had to change to cope. Obviously, the authors here will have processed these messages, and the results will not be exactly as farmers see land degradation. Nevertheless, the principles of field observability and farmer relevance will be maintained throughout the rest of this book in deciding what to include and what to exclude.

Advantages of the Farmer-perspective Approach

There are three main advantages of adopting a farmer-perspective approach to land degradation assessment. First, measurements are far more realistic regarding actual field-level processes. Second, assessments utilize the integrated view of the ultimate client for the work, the farmer. Third, results provide a far more practical view of the types of interventions that might be accepted by land users.

Realism. The problem with most techniques of scientific monitoring of degradation processes is that they intervene in the process itself. Measurements may simply reflect the intervention rather than the process in its real field setting. Runoff plot results, for example, are affected by the boundary barriers of the plot. These barriers alter the erosion processes by decreasing overland flow of water from upslope but increasing the channelization of flow along the plot boundary. Even a simple erosion pin (a long thin stake forced into the ground, against which lowering of the level of topsoil can be measured) has its problems. The insertion of the stake may crack the soil, altering the local hydrology and resistance to erosion. The stake itself affects runoff around it, possibly causing down-slope eddies in the water current. Stakes are also very likely to be interfered with by children and inquisitive cattle. Accuracy of measurement of very small changes in ground surface is extremely difficult, especially considering that 1 tonne of soil loss per hectare is equivalent to much less than

0.1mm lowering. So, small errors introduced by the measurement technique have potentially large implications for the measurement result.

Conversely, most of the field techniques in this handbook rely on the results of processes that have not been altered by the technique of monitoring. So, accumulations of sediment against a barrier such as a boundary wall of a field are 'real' accumulations that would have occurred whether or not an observer were interested in measuring them. In addition, measuring the height of a mound of soil protected by a tree, relative to the general level of the soil surface influenced by erosion since the tree started to grow, is a 'real' difference that is impossible to ascribe to inaccuracies introduced by the technique of measurement. There may be other explanations for the tree mound (see 'Health' Warnings below) but these are no more serious than alternative explanations in other more interventionist techniques.

Realism is also enhanced by simple field techniques in that indicators often used by farmers are being employed. The pedestals under small stones and the existence of coarse sandy and gravelly deposits in fields are both frequently identified by farmers as the result of rain-wash (see Plate 3). Stunted growth of plants and markings on leaves are also recognized by farmers and ascribed to deterioration in the quality of the soil. If the field assessor uses the same indicators that the land user employs, then they talk the same 'language'. The shared understanding leads to greater realism and cooperation.

Integration. The results derived from field assessments tend to integrate a wide variety of processes of land degradation. This is most evident in changes in soil productivity as measured by farmers' assessments of historical yield. Many scientists may see this as a disadvantage, covering up the causative influences on yield reduction. Yields are a product not only of soil erosion, but of past and current management, seed sources, climate, pests and general vagaries of nature. However, land degradation is a very broad concept, including not only attributes of the physical environment but also the way in which the environment is managed and how nature reacts to human land use. So, integration is essential if the researcher is to present the outcome of a set of processes that farmers really face. The scientific method of deconstructing natural processes into their singular elements for study, and then reassembling them to regain complex reality, has dubious validity in ecological systems where it is the interactions between components that are far more influential.

Take the example of how vegetation controls soil degradation. Directly, vegetation introduces organic matter into soil, which renders the soil less erodible. But, indirectly, and of far greater universal importance, is the way that vegetation cover intercepts raindrops. The energy of the drops is dissipated in the structure of the plant, rather than being used to dislodge soil particles. If soil remains undetached, then it cannot be transported. These interaction effects are vital to capture if accurate assessment of the severity of degradation is to be made.

Practicality. Of most significance, however, is that farmer-perspective assessments are more practical. They bring together the long experience of the farmer in using the field – experience that could not possibly have been accumulated by the researcher as an occasional visitor. They capture the observations, understandings and reactions of the farmer – insights gained over years, possibly over generations. The researcher can also learn much about how farmers respond to the effects of land degradation from in-field experimentation by farmers. Farmers experiment in many areas – they try new varieties, vary planting dates and test different fertility treatments and conservation measures. Farmer-perspective assessments, therefore, access a storehouse of knowledge that could not be gained by any other means.

Practicality also extends to the application and use of the results. If, for example, the farmer has been involved in collection and processing of field data on land degradation, then ownership of the results and empowerment of the land user are far more clearly identified with the farmer rather than the researcher. This participatory element has been found to be essential in most rural development work. Furthermore, results of land degradation assessments will be much more relevant to the issues facing land users. Change in soil productivity that affects future yields is a constant concern to many marginal land users. So, land degradation assessments which use yield as the indicator variable will much more closely relate to farmers' priorities and be much more likely to develop solutions which combat land degradation through yield-enhancing measures.

A further practical attribute of field-level farmer-perspective assessments is that they are quick and simple. Many more observations can be accomplished in a short time than through the more complex procedures of standard monitoring. In a field visit along with a farmer using the sorts of techniques advocated in this handbook, a far greater range of types of measurement can be made than through standard, narrowly focused approaches. Having the possibility of many data points enables a much better sampling of the enormous number of permutations of field types, management regimes, crops and land uses. Having several different types of measurements enables better cross-checking of results and the search for consistent trends in land degradation.

'Health' Warnings

 Farmer-perspective assessment is not without its limitations. It is not a panacea for all the ills of the land. Problems relate to **accuracy**, **extrapolation** and **reliability**.

The **accuracy** of individual observations is often compromised. Any single measurement derived by an assessor on hands and knees while clinging to a steep slope is unlikely to be as accurate as a measurement in the peace of the laboratory. Using a ruler marked in millimetres to measure the effect of a process that is significant at an order level of one less (ie 0.1mm) inevitably introduces inaccuracies. Inaccuracy also means that results are difficult to replicate from place to place. The very nature of land degradation often means it is spatially discontinuous: a gully, for example, punctuates a landscape. These problems are partially compensated by undertaking large numbers of such measurements. Triangulation (described below) and combinations of indicators described in Chapters 4 and 5 also serve to reduce the problems of inaccuracy.

Because farmer-perspective assessments tend to integrate the effect of a variety of often-unknown processes, it is very difficult to **extrapolate** the results to unmeasured conditions. If, for example, it were known that aluminium toxicity causes yield declines after a crop that allows high erosion, then these same conditions would be likely to prevail at another broadly similar geographical location. But farmer-perspective assessments usually contain only limited information on causative relationships. Hence, extrapolation to other places is problematic. Partly, this can be overcome by undertaking parallel investigations into the scientific rationality of farmers' techniques. In one study in semi-arid Kenya, for example, farmers used trashlines (barriers of weeds placed along the contour) as a conservation measure, even though the specialists were not recommending them. At the same time they routinely ignored the advice to build large-scale terraces. Further investigation revealed that the farmers were absolutely right! Terraces would have had a negative effect on their farm economy. Trashlines were far more effective in maintaining soil fertility levels and soil humidity, and were extremely low-cost to construct. In another example, many different conservation strategies (such as compost mounds or deep ditches) can be observed in the sweet potato-growing highlands in Papua New Guinea. Investigation has revealed that the main function of each of these diverse strategies is to aerate the soil, since sweet potato is particularly sensitive to wet soil. It is these sorts of insights that enable farmers' knowledge to be extrapolated. Successful, locally developed agricultural practices, which protect against land degradation, can be identified and disseminated to farmers in similar circumstances.

Finally, it has been claimed that farmer-perspective assessments are less **reliable**. It is true that many means of controlling reliability are unavailable to the researcher. How does one know the farmer is telling the truth, for example? How can one be sure that a mound of soil at the base of a tree is a residual left behind after unprotected soil around the tree has been eroded, rather than, say, an old termite mound that happened to occur under the tree? How can such a wide variety of techniques be used from field to field, without more consistent guidance as to applicability and relevance? Part of the answer to such problems lies in the social science technique of 'triangulation'. Triangulation is the use of several different methods or sources of information to gain a consensus view of a situation, such as the status of land degradation. Obviously, the different methods give different representations of absolute levels of land degradation. But their combined message, if in agreement, gives a powerful conclusion; far more powerful than the results of just one measurement technique. (See also Box 6.1.)

What is Included Here under 'Land Degradation'?

Land degradation is a composite term, which is explained fully in Chapter 2. However, there is considerable confusion over what is included within the term and how best to represent it in practical, field terms.

The approach adopted in this book is to view land degradation as an 'umbrella' term, covering the many ways in which the quality and productivity of land may diminish from the

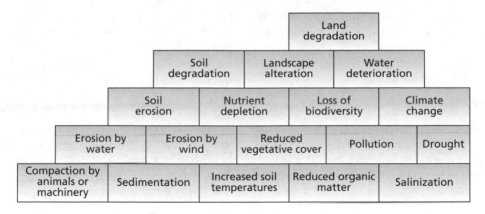

Land degradation consists of many components, each of which interlocks with many other components

Figure 1.1 *The Land Degradation Wall*

point of view of the land user (and of society at large). It therefore includes changes to soil quality, the reduction in available water, the diminution of vegetation sources and of biological diversity, and the many other ways in which the overall integrity of land is challenged by inappropriate use (see Figure 1.1). Land degradation also includes many urban and industrial problems, such as pollution, mine tailings, smog and waste dumping.

Clearly, to make assessment of land degradation viable, indicators of its process and effect have to be used. These indicators may be drawn from any aspect of how the quality of land degrades. Since there is much interlinkage between the various types and manifestations of land degradation, indicators give us a powerful tool for overall assessments. For example, a reduction in vegetation cover through deforestation will almost always be accompanied by soil erosion, sedimentation of lower slopes and increased surface runoff. Excessive runoff from agricultural lands often goes along with the pollution of water bodies and downstream flooding. An indicator such as 'Armour Layer' (see Chapter 4) can be used, therefore, not only

to assess in-field water erosion but also – with careful observation – to assess off-site aspects of land degradation.

This handbook has deliberately concentrated on those indicators of closest relevance to farmers and land users. First, they concentrate on **soil degradation**. This is one manifestation of land degradation that focuses on soil quality and soil productivity. Although soil degradation is only one aspect of land degradation, variables of its progress can be used as indicators of land degradation. Soil degradation, itself, is also conceptually rather wide and difficult to accommodate in a few simple measures. **Soil erosion by water** is, for most landscapes, the most common way in which soil degradation occurs. Again, there is considerable linkage between erosion and other types of degradation. 'Nutrient mining', or the depletion of soil nutrients through taking more nutrients away in the harvested crops than are returned, is less visible but is a common cause of soil fertility decline. Soil erosion by water is one of the driving forces for such depletion of nutrients. An eroded soil will almost always have less organic matter (biological soil degradation), increased bulk density (physical soil degradation) and other problems such as waterlogging. Salinity and sodicity are more restricted, but even they commonly occur along with other aspects of soil degradation.

Since soil is the medium through which many, if not most, changes in landscape status occur, it is reasonable to use it as the focus for assessing land degradation. And since soil erosion by water is the most visible way in which land degradation affects the direct production of land users, this book has deliberately taken evidence of soil erosion as the main set of indicators of the seriousness of land degradation. This is not done to imply that soil erosion is the only (or even the single most important) evidence of land degradation which affects farmers; but there are many experiences to indicate that soil erosion acts as the single best proxy for most of the other aspects of degradation.

What is Land Degradation?

Definition

Land degradation is a composite term; it has no single readily identifiable feature, but instead describes how one or more of the land resources (soil, water, vegetation, rocks, air, climate, relief) has changed for the worse. The change may prevail only over the short term, with the degraded resource recovering quickly. Or it may be the precursor of a lengthy downward spiral of degradation, causing a long-term, permanent change in the status of the resource. Take, for example, a landslide. It is often viewed as an example of land degradation in action: it changes the features of the land, sweeps away productive resources, causes destruction of houses and disrupts activities. The landslide creates massive and disastrous change. In the longer term, however, the area of a landslide may regain its productivity. In places such as Jamaica and Papua New Guinea, old landslide scars are noted for supporting better crops and more intensive agricultural possibilities than the adjacent land not affected by landslides, especially when the new soil is derived from less weathered rock materials, such as calcareous mudstones. The irony of degradation leading to long-term improvement of the natural resource base has not been lost on local people. They are quick to exploit the new opportunities. So, land degradation is far from being a simple process, with clear outcomes. This complexity needs to be appreciated by the field assessor, before any attempt is made either to define land degradation or to measure it.

Land degradation is generally defined as the temporary or permanent decline in the productive capacity of the land (UN/FAO definition). Another definition describes it as 'the aggregate diminution of the productive potential of the land, including its major uses (rain-fed, arable, irrigated, rangeland, forest), its farming systems (eg smallholder subsistence) and its value as an economic resource'. This link between degradation (which is often caused by land use practices) and its effect on land use is central to nearly all published definitions of land degradation. The emphasis on land, rather than soil, broadens the focus to include natural resources such as climate, water, landforms and vegetation. The productivity of grassland and forest resources, in addition to that of cropland, is embodied in this definition. Other definitions differentiate between reversible and irreversible land degradation. While the terms are used here, the degree of reversibility is not a particularly useful measure; given sufficient time all degradation can be reversed, as illustrated by the landslide example above. So, reversibility depends on whose perspective is being assessed and what timescale is envisaged (see Box 2.1). While soil degradation is recognized as a major

---

BOX 2.1 REVERSING LAND DEGRADATION THROUGH SOIL FORMATION

Soils form through natural processes of chemical and physical breakdown of rocks, along with the addition of biota that start the cycling of nutrients. Remove the soil by erosion and soil formation must start all over again. How long does this aspect of land degradation take to be reversed?

Experiments in Zimbabwe have shown that even under relatively humid conditions, soils on the dominant geology of the country – granite – form only at a rate of between 400 and 800 kg/ha/yr. This equates to much less than an annual increase in depth of 0.1mm of soil. Enough soil to give a reasonable rooting depth for plants (50cm) would take more than 5000 years to form. These figures provide a stark illustration of the importance of conserving land resources rather than letting them degrade and waiting for nature to reverse the damage.

---

aspect of land degradation, other processes which affect the productive capacity of cropland, rangeland and forests, such as lowering of the water table and deforestation, are captured by the concept of land degradation.

While most definitions see land degradation in its relation to the biophysical environment, it is important to appreciate that degradation is socially constructed. The social construction may be seen in two ways. First, in relation to the use of the land resource by society, a change in quality has a social construction. The more that people rely on a resource that is degrading, the greater is the disruption to society arising from the degradation. Therefore, the greater is the perceived degradation. A farmer whose only source of livelihood is the intrinsic quality of the land resources of the farm suffers more than another farmer who can augment income from employment elsewhere. Similarly, a land resource that is essential for survival in the sense that there is no other livelihood resource will usually be protected carefully. Societies whose only source of living is their land have usually developed some of the best land management practices. The *ngoro* (or *Matengo*) pits of Mbinga District, southern Tanzania, are a remarkable case in point.

The people of these isolated highlands have developed an intricate cross-ridging land management technique, using buried weeds and grass as a mulch for fertility, and the pits so formed as a way of conserving water on steep slopes (see Plate 10). Second, social construction must be seen in terms of who does the perceiving. One section of society may put a completely different construction on evidence of land degradation than another part of society. The people of Kissidougou provide an excellent example (Box 2.2), to the extent that the different social constructions come to opposite conclusions on land degradation. Definitions of land degradation must, therefore, incorporate the relationship of change in biophysical quality of the land to the effect it has on society, economy, politics and humanity; the attributes of sensitivity and resilience of land, introduced later in this chapter, will help to capture this crucial relationship.

Land degradation is, however, difficult to grasp in its totality. The 'productive capacity of land' cannot be assessed simply by any single measure. Therefore, we have to use indicators of land degradation. Indicators are variables which may show that land degradation has taken place; they are not necessarily the actual degradation itself. The piling up of sediment against a

---

BOX 2.2 LAND DEGRADATION AS A SOCIAL CONSTRUCTION: AN EXAMPLE

Since the late 19th century the authorities (first the French and later national) assumed that the forest islands surrounding the villages in Kissidougou, Guinée, were the remnants of a once extensive forest. The assumption was that the forest had been gradually destroyed as land users converted it to agricultural use. This 'savannization' was identified as continuing degradation of the environment, and led to efforts to rehabilitate the perceived degradation.

This interpretation of a degraded and degrading landscape took no account of the opposite view of savannization held by the local communities living in the villages in the forest islands. They understood the forest areas to have been created, out of the savanna, by the villagers themselves, or by their ancestors.

So we have two social constructions of what is happening to the land. The colonial authorities found it useful to see the forest islands as evidence of degradation – it served to strengthen the need for them to govern an 'unruly peasantry'. Local people found it useful to see the forest as evidence of their improvement of the land – it served to support their exclusive tenure of the land.

The fascinating postscript to these social constructions arises from the work of James Fairhead and Melissa Leach. Evidence of vegetation change over time largely supports the construction of events of the local people. In Kissidougou we have a savanna partly filled and progressively filling with forests.

Source: Misreading the African Landscape: Society and Ecology in a Forest–Savanna Mosaic, James Fairhead and Melissa Leach

---

downslope barrier may be an 'indicator' that land degradation is occurring upslope. Similarly, decline in yields of a crop may be an indicator that soil quality has changed, which in turn may indicate that soil and land degradation are also occurring. The condition of the soil is one of the best indicators of land degradation. The soil integrates a variety of important processes involving vegetation growth, overland flow of water, infiltration, land use and land management. Soil degradation is, in itself, an indicator of land degradation. But, in the field, further variables are used as indicators of the occurrence of soil degradation. This chapter and much of the rest of this handbook will, therefore, dwell primarily on the use of evidence from the soil (mainly soil degradation) and from plants growing on the soil (soil productivity). This first is a measure of change in biophysical quality; the second assesses an aspect of impact on society. Both are essential.

Types of soil degradation include:

1 **Soil erosion by water:** the removal of soil particles by the action of water. Usually seen as sheet erosion (a more or less uniform removal of a thin layer of topsoil), rill erosion (small channels in the field) or gully erosion (large channels, similar to incised rivers). One important feature of soil erosion by water is the selective removal of the finer and more fertile fraction of the soil.

2 **Soil erosion by wind:** the removal of soil particles by wind action. Usually this is sheet erosion, where soil is removed in thin layers, but some-

times the effect of the wind can carve out hollows and other features. Wind erosion most easily occurs with fine to medium size sand particles.

3   **Soil fertility decline:** the degradation of soil physical, biological and chemical properties. Erosion leads to reduced soil productivity, as do:

- reduction in soil organic matter, with associated decline in soil biological activity;
- degradation of soil physical properties as a result of reduced organic matter (structure, aeration and water-holding capacity may be affected);
- changes in soil nutrient content leading to deficiencies or toxic levels of nutrients essential for healthy plant growth;
- build up of toxic substances, for example pollution or incorrect application of fertilizers.

4   **Waterlogging:** caused by a rise in groundwater close to the soil surface or inadequate drainage of surface water, often resulting from poor irrigation management. As a result of waterlogging, water saturates the root zone, leading to oxygen deficiency.

5   **Increase in salts:** this could either be salinization, an increase in salt in the soil water solution, or sodication, an increase of sodium cations (Na^+) on the soil particles. Salinization often occurs in conjunction with poor irrigation management. Mostly, sodication tends to occur naturally. Areas where the water table fluctuates may be prone to sodication.

6   **Sedimentation or 'soil burial':** this may occur through flooding, where fertile soil is buried under less fertile sediments; or wind blows, where sand inundates grazing lands; or catastrophic events such as volcanic eruptions.

In addition to these principal types of soil degradation, other common types of land degradation include:

7   **Lowering of the water table:** this usually occurs where extraction of groundwater has exceeded the natural recharge capacity of the water table.

8   **Loss of vegetation cover:** vegetation is important in many ways. It protects the soil from erosion by wind and water and it provides organic material to maintain levels of nutrients essential for healthy plant growth. Plant roots help to maintain soil structure and facilitate water infiltration.

9   **Increased stoniness and rock cover of the land:** this is usually associated with extreme levels of soil erosion causing exhumation of stones and rock.

Examples of apparently severe degradation are illustrated in Plates 4 to 8. Note that in each case the bareness of the soil and the relative lack of vegetation are common features, although the process leading to degradation may be quite different.

Although the foregoing list neatly breaks down the components of soil degradation by cause, very often these agents of degradation act together. For example, strong winds often occur at the front of a storm, thus wind erosion and water erosion may result from the same event. A sodic soil (Plate 6) where sodium cations have accumulated is also the single most susceptible soil type to water erosion in the tropics. Steep hill lands suffering water erosion (Plate 7) are associated with lower slope sedimentation. Additionally, a soil that has suffered some form of degradation, such as burning of vegetation (Plate 8), is more likely to be further degraded than another soil similar in all respects except for the level of degradation. An example of how some degradation tends to lead to more degradation is the central role of the carbon or organic matter cycle in giving the land greater resilience. One well-accepted indicator of increased erodibility is the level of soil organic matter. Where the organic matter content of a soil falls below 2 per cent, the soil is more prone to erosion, because soil aggregates are less strong and individual particles are more likely to be dislodged.

Some environments are naturally more at risk to land degradation than others (see Box 2.3). Factors such as steep slopes, high intensity rainfall and reduced soil organic matter influence the likelihood of the occurrence of degradation. Identification of these factors allows land users to implement techniques that safeguard against loss of productivity. Management practices also exert a significant influence on the susceptibility of a landscape to degradation. Extensive and poorly managed land use systems are more likely to degrade than intensive, intricately managed plots. Multi-storey homestead plots of the humid

tropics, as typified by the Kandy home gardens of Sri Lanka, have resisted degradation for centuries, whereas extensive cultivation for cereals of the former Maasai rangelands of East Africa has led to widespread sheet erosion.

Milder forms of land degradation can be reversed by changes in land management techniques, but more serious forms of degradation (such as salinity) may be extremely expensive to reverse or may be, for practical purposes, irreversible. Box 2.4 discusses how land users may compensate for land degradation. If these applications of technology are economically and ecologically sustainable, then reversal of land degradation has been achieved. But if technology buys only time (eg added fertilizer depletes quickly; typically more and more chemicals need to be added each year to achieve the same level of yields) then the land user is just 'hiding' land degradation. Soil erosion, when serious and prolonged, is effectively irreversible because, in most circumstances, the rate of soil formation is

---

BOX 2.3 'AT-RISK ENVIRONMENTS': FLOOD-PRONE AREAS IN PERU

Land degradation occurs under a wide variety of conditions and circumstances. Nevertheless, some environments are more at risk of degradation. This risk of degradation affects how people manage their biophysical environment and also how their environment affects them. A good example comes from the People, Land Management and Environmental Change (PLEC) project sites in the Peruvian Amazon, which are subject to two different types of flooding.

The first occurs in coastal regions as a result of inundations from the sea. The second type of flooding is the annual increase in river levels in Amazonia which results in flooding of the land along the riverbanks. Much of the agricultural production in Peruvian Amazonia takes place along the riverbanks where the level of soil fertility is very high. Such annual flooding is part of the agricultural cycle and, as such, is planned for by local people.

The flood level is critical in determining the effect of flooding. Exceptionally high flood levels can lead to reduced pest and weed levels, improved hunting and better fishing in the next year, but if the higher areas are also flooded, crops may be destroyed and food scarcity may ensue. Very high levels of sedimentation, particularly of sand, can change the landscape completely. Fast flowing floods may result in severe riverbank erosion and the loss of valuable agricultural land close to the river. On the other hand, when the flood level is low the staple crops grown in the relatively high areas are not endangered but pests survive and, if there is little sedimentation, fertility replenishment may be poor.

Source: Miguel Pinedo, PLEC-Peru Cluster Leader, personal correspondence

so slow (see Box 2.1). In moist, warm climates, formation of just a few centimetres of soil may take thousands of years, and in cold, dry climates it can take even longer. Soil loss through erosion happens far faster: up to 300 times faster where the ground is bare.

Soil erosion is the most widely recognized and most common form of land degradation and, therefore, a major cause of falling productivity. However, since the effects of soil loss vary depending on the underlying soil type, soil loss, by itself, is not an appropriate proxy measure for productivity decline. For example, a loss of 1mm from a soil in which the nutrients are concentrated close to the surface (eg a Luvisol; see Appendix V) will show a greater impact on productivity than the same level of soil loss from a soil in which the nutrients are more widely distributed (eg a Vertisol; see Appendix V).

In Table 2.1, estimates of soil loss rates under different types of land management are summarized. These rates are based on typical soil loss plot data from Zimbabwe. They demonstrate the huge impact that manipulation of the environment by humans can have on rates of soil erosion. The rate of soil loss from bare soil is 250 times that from areas covered by natural forest. Even the rate of soil loss from a well-managed cropping system is ten times greater than that from under natural ground cover. Natural forest best represents the situation where soil loss is in approximate balance with the rate of soil formation.

Although land degradation is defined by reference to productivity, its effects may include diminished food security, reduced calorie intake, economic stresses and loss of biodiversity. These consequences concern rural land users greatly, and will be addressed wherever possible in the following chapters as an important part of field assessment of land degradation.

BOX 2.4 HOW TO 'HIDE' LAND DEGRADATION

There are a number of ways to compensate for land degradation so that production may continue – but at a cost:

Shhhhh

- If soil nutrients are depleted through erosion, leaching into groundwater or excessive removal in crops, they can be replaced through the application of manufactured fertilizer
- If the water-holding capacity of the soil is reduced due to depletion of organic matter or failure to renew natural losses, this can be compensated for by irrigation
- If soil structure degenerates and hard pans form under heavy machinery, new deep rippers can be used to break up the pan and aerate the soil

In all cases, the land is still degraded. Its productive potential, by virtue of intrinsic soil quality, has been diminished. But the effects of land degradation can be hidden, and production can be maintained through inputs of technology. For example:

- New seed varieties to give better yields on degraded soils
- Irrigation to replace lost plant-available water
- Machinery to correct for soil physical damage
- Fertilizer to replace lost nutrients
- Glasshouses and hydroponic technology to replace land, climate and soil

Table 2.1 *Typical Relative Measures of Soil Loss According to Land Use*

Land use	Soil Loss Rate (tonnes/ha/yr)
Bare soil	125.0
Annual crops – poor management on infertile soil	50.0
Annual cropping – standard management	10.0
Annual cropping – good management	5.0
Perennial crops – little disturbance	2.0
Natural forest	0.5

Source: Soil loss plot results from Zimbabwe, on a 9 per cent slope

Causes of Land Degradation

Although degradation processes do occur without interference by man, these are broadly at a rate which is in balance with the rate of natural rehabilitation. So, for example, water erosion under natural forest corresponds with the subsoil formation rate. Accelerated land degradation is most commonly caused as a result of human intervention in the environment. The effects of this intervention are determined by the natural landscape. The most frequently recognized main causes of land degradation include:

- overgrazing of rangeland;
- over-cultivation of cropland;
- waterlogging and salinization of irrigated land;
- deforestation;
- pollution and industrial causes.

Within these broad categories a wide variety of individual causes are incorporated. These causes may include the conversion of unsuitable, low potential land to agriculture, the failure to undertake soil conserving measures in areas at risk of degradation and the removal of all crop residues, resulting in 'soil mining' (ie extraction of nutrients at a rate greater than resupply). They are surrounded by social and economic conditions that encourage land users to overgraze, over-cultivate, deforest or pollute. These are considered in the following chapter.

Biophysical conditions also affect an understanding of causes of land degradation. The effect of a land degrading process differs depending on the inherent characteristics of the land, specifically soil type, slope, vegetation and climate. Thus an activity that in one place is not degrading may in another place cause land degradation because of different soil characteristics, topography, climatic conditions or other circumstances. So, equally erosive rainstorms occurring above different soil types will result in different rates of soil loss. It follows that the identification of the causes of land degradation must recognize the interactions between different elements in the landscape which affect degradation and also the site-specificity of degradation.

It is possible to distinguish between two types of land degrading actions. The first is unsustainable land use. This refers to a system of land use that is wholly inappropriate for a particular environment. It is unsustainable in the sense that, unless reversed, this land use or indeed any other could not be continued into the future. Unsustainability has the implication of being irreversibly degrading. Many 'badlands' (extremely bare, devegetated and eroded slopes) are effectively irreversibly degraded. However, a large input of technology could start a rehabilitation process, if enough time and resources were to be devoted. Usually, this is uneconomic. Second, inappropriate land management techniques also cause land degradation, but this degradation may be halted (and possibly reversed) if appropriate management techniques are applied.

In considering the cause of land degradation, care must be exercised in using terms such as 'unsustainable' and 'inappropriate'. There is a danger of circular reasoning: the use of a hypothesis to reach a result that therefore proves the hypothesis. Assuming that land degradation has been caused by inappropriate land management tends to reinforce the view that land management practices are inappropriate. This type of reasoning is common among those who blame land users for their unwise use of land. The neat circularity of the argument not only labels land users as foolish but also diverts attention from other causes. Political ecology teaches us that there may indeed be a **proximate** cause in, for example, a farmer cultivating a very steep slope without conservation measures. However, there are **intermediate** causes for the degradation in why the farmer is forced to do so – for example, lack of more suitable land – and **ultimate** causes in the structure of society, economy and power relations that explain why that farmer has been unable to get suitable land. 'Cause' is therefore a loaded word. Searching for causality is another way in which perceptions of land users are distinguishably different from those of professionals. To re-emphasize the point, it is the farmers' perceptions that must be prioritized in the mind of the field assessor of land degradation.

Farmers' Concerns

In order to promote farmers' concerns a distinction is made between productivity, which is defined as the inherent potential of a land system to produce crop yields, and production, which is defined as the actual yield levels achieved by farmers. Farmers are concerned with both, to different degrees and under different circumstances. Land degradation may reduce the inherent productivity of a system, but production levels may be unaffected, or may increase as a result of compensating action being taken by the land user (for example, the application of fertilizer). Land management practices may not exploit the full potential productivity of the land.

Land degradation, if defined as a loss in productivity, is closely aligned with the interests of farmers, whose major concern is the yield that they can achieve in the future from their lands. It is not, however, their only concern. Although current harvest potential is critical to most farming decisions, farmers will often take a longer-term approach to land productivity. Farmers may wish to control degradation to impress the neighbours or curry favour with government agents. Farming activities can trigger or exacerbate land degradation, storing up future problems for land users. Consider two extremes that reflect the differing strategic concerns of a farmer. One farmer may expend much effort in creating an intricate system of farming with agroforestry, bench terraces, stall feeding of animals, composting and so on. The increased effort may not immediately be rewarded in direct financial terms by a compensating increase in production. Future production or soil productivity is an unknown. Why does the farmer do it? Farmers' legitimate concerns also include power, prestige, influence, access to extension and credit services or any number of local socio-political reasons. Through these means, productivity in a broad sense may be enhanced. Power, for example, could lead to a better livelihood or certainly a more satisfying existence. Another farmer may 'mine' his soil, being concerned with immediate production to pay school fees, invest in a brother's business or buy another wife. The land degradation directly caused by concerns for production is really driven by concerns far from production. Those latter concerns may, however, enhance long-term productivity through, for example, the younger wife's active contribution to labour. So, in the first case, a potential loss in productivity is averted by the concerns of the farmer for a more secure livelihood in the widest sense. In the second case, the land degradation caused by the immediate concern for production brings better productivity in the long-term. These are complex issues, more fully explored in the following chapter. For the moment, however, the central importance of farmers' concerns must be firmly noted by field assessors in their understanding of why degradation is occurring.

Sensitivity and Resilience

Sensitivity and resilience are measures of the vulnerability of a landscape to degradation. It is vital that the field assessor understands these terms since they incorporate evidence of the 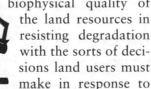 biophysical quality of the land resources in resisting degradation with the sorts of decisions land users must make in response to the changing conditions. These two factors combine to explain the degree of vulnerability.

Sensitivity is the degree to which a land system undergoes change due to natural forces, human intervention or a combination of both. Some places are more likely to be sensitive to change – for example, steep slopes, areas of intense rainfall or highly erodible soils. These places are subject to natural hazards that make them sensitive to change. Human intervention in these systems can result in dramatic alterations. Sensitivity to change can arise as a result of human intervention; for example, in a natural state forested hillsides may be difficult to degrade, but once converted to farmland degradation may occur more easily.

Resilience is the property that allows a land system to absorb and utilize change, including resistance to a shock. It refers to the ability of a system to return to its pre-altered state following change. The natural resilience of an environment may be enhanced by the diversity of the land management practices adopted by land users. Degraded land is less resilient than undegraded land. It is less able to recover from further shocks, such as drought, leading to further degradation.

Table 2.2 summarizes the relationship between resilience and sensitivity of ecosystems. Where a landscape is susceptible to change (high sensitivity) the risk of

Table 2.2 *Sensitivity and Resilience*

	Sensitivity	
	High	Low
High resilience	Easy to degrade Easy to restore capability	Hard to degrade Easy to restore capability
Low resilience	Easy to degrade Hard to restore capability	Hard to degrade Hard to restore capability

Note: This matrix shows the extremes of both sensitivity and resilience. In practice, most land systems are likely to fall somewhere between the high and low points

degradation is affected by the resilience of that landscape – high resilience lessens the danger of serious degradation, whereas low resilience indicates that changes are not likely to be easily reversible and may even be permanent. Land systems that exhibit high resilience are likely to return to their previous stable state following disruption, whereas systems with low resilience are more likely to be permanently altered by such disruption.

Advance recognition of the sensitivity and resilience of a land system should influence land use decisions, thereby reducing the risk of permanent degradation to the system. Similarly, the sensitivity and resilience of specific soil types also alerts the field assessor to the risk of permanent or temporary soil degradation. For example, an iron-rich but highly weathered and acid Ferralsol (see Appendix V) of the humid tropics has a low sensitivity to degradation as well as low resilience. So, once it has been degraded (which is difficult to do in a physical sense), then it is almost impossible to bring back to a productive state. Contrast this with a Phaeozem (see Appendix V) that has high organic matter and an excellent structure. Under good management Phaeozems give consistently high yields, but with poor management they degrade very quickly. This high sensitivity is moderated somewhat by a high resilience because, using organic methods, the soil can be rehabilitated fairly quickly.

What Characteristics Contribute to Sensitivity and Resilience?

The factors that affect sensitivity and resilience of an environment are the inherent characteristics of that environment (ie soil properties such as nutrient reserves, soil structure, micro-aggregates and soil depth, topography, climate, etc) and the human element, in the form of land use and management practices. The salient features affecting sensitivity and resilience will vary from place to place.

So, with regard to aspects of land degradation, sensitivity refers to how easy it is to degrade the land, and resilience to how easy it is to restore the land. Some combinations of factors that may influence the sensitivity and resilience of land systems are suggested in Table 2.3 overleaf, with highlighting of those combinations that might be especially vulnerable. This matrix illustrates how different combinations of factors affect the sensitivity and resilience of a system in different ways. For example, the sensitivity of a Vertisol is low (ie hard to degrade)

Table 2.3 *Examples of Resilience and Sensitivity under Different Conditions*

Factors Related to Land Use and Users	Factors Related to Intrinsic Properties of Soil, Climate and Landscape								
	Luvisol	Vertisol	Sodic soils	Low initial soil organic matter	Intense rainfall	Drought-prone	Strongly seasonal rainfall regime	Low plains	Steep slopes
Intensive home gardening with organic inputs	High S High R	Low S High R	High S High R	Int S High R	High S High R	Int S High R	Int S High R	Low S High R	High S High R
High artificial input field crops	High S Int R	Low S Int R	High S Int R	Int S Int R	High S Int R	Int S Int R	Int S Int R	Low S Int R	High S Int R
Extensive conventional ploughing with poor management	High S Low R	Low S Low R	High S Low R	Int S Low R	High S Low R	Int S Low R	Int S Low R	Low S Low R	High S Low R
Shifting cultivation – long cycle	High S Int R	Low S Int R	High S Int R	Int S Int R	High S Int R	Int S Int R	Int S Int R	Low S Int R	High S Int R
Short cycle bush fallowing	High S Low R	Low S Low R	High S Low R	Int S Low R	High S Low R	Int S Low R	Int S Low R	Low S Low R	High S Low R
Perennial tree and bush crops	High S High R	Low S High R	High S High R	Int S High R	High S High R	Int S High R	Int S High R	Low S High R	High S High R
High input grazing	High S High R	Low S High R	High S High R	Int S High R	High S High R	Int S High R	Int S High R	Low S High R	High S High R
Extensive rangeland	High S Low R	Low S Low R	High S Low R	Int S Low R	High S Low R	Int S Low R	Int S Low R	Low S Low R	High S Low R

Key: S = Sensitivity   R = Resilience   Int = Intermediate

Vulnerable Combinations   High S Int R

Highly Vulnerable Combinations   High S Low R

because the abundant clays and organic matter, as well as its usual low-lying position in the landscape, make it difficult to degrade. A Luvisol, on the other hand, with its loose sandy texture, limited reserves of nutrients and low organic matter status, is easy and quick to degrade.

Combining these mainly naturally inherited qualities of sensitivity with land use factors of resilience enable an overall assessment of the degree of **vulnerability** to land degradation to be undertaken. A Luvisol undergoing extensive conventional ploughing up and down the slope not only degrades extremely quickly, but is also very difficult to restore (low resilience) – ie it is extremely vulnerable. A Vertisol used for intensive organic gardening is right at the other end of the scale of vulnerability.

Scientific Interpretation of Degradation Compared with Land Users' Perceptions

Often, the views of scientists and the opinions of land users do not coincide. As discussed in Chapter 1, the land user's concern is most likely to be productivity. Thus, the existence of land degradation, of itself, is unlikely to be a cause of much concern, unless it has an adverse effect on productivity. What may be seen by a scientist as a potentially degrading situation may have a different significance for farmers. Some examples of the interpretation of land degrading processes by both scientists and farmers are set out in Table 2.4. These represent two extremes – most often there is overlap between the understanding of the land user and the scientist. It is important to appreciate how land users perceive processes of land degradation if discussions about land degradation and preventive measures are to have any relevance for them.

Take the example of rills seen in the field. They are obvious evidence of land degradation to the professional, and they will be used as one of the field indicators of soil loss in Chapter 4. Local farmers will also see the rills, but view them differently. In an extreme case, Sri Lankan hill farmers have extolled the virtues of rills in giving natural drainage to their steep vegetable plots, providing a useful line in which to place weeds and enabling ready access into the field for harvesting. Of course, many farmers would also see the negative aspects of rills: washing out of seeds, undermining crops along the line of the rill and requiring extra labour to obliterate for the next season. But the important point is that the scientific interpretation of a rill's seriousness will usually be different from the land user's interpretation.

The accurate measurement of soil loss through erosion is of legitimate interest to scientists. However, land users are generally more concerned about the effects of erosion than the absolute amount of soil loss. This handbook focuses on quick methods of measuring soil loss and of assessing negative effects on the productivity of the land. It does not seek to describe procedures that will give results which would meet the rigours of scientific measurement. Instead, the aim is to provide extension workers and land users with accessible techniques that will provide a sufficient basis for planning future actions to protect and increase the productivity of the land.

Table 2.4 *Two Extremes in the Interpretation of Outcomes of Land Degradation Evidence*

Scientific Interpretation	Process	Land Users' Interpretation
High erosivity and potential soil erosion	Heavy rainfall	Damage to crops. But also benefit to soil and planting opportunity
Loss of finer soil particles through water or wind erosion	Stones on the soil surface	Soil formation (Burungee people, Dodoma region, Tanzania)
Increased risk of soil loss through water erosion	Planting crops up and down steep slopes rather than across	Protection of crop from waterlogging and/or wind damage
Severe erosion and abuse of catchment	Deep gullies	Livestock fatalities and loss of roads/bridges
Severe short term erosion, indicating need for better cover	Rills	Useful local drainage channels to prevent waterlogging and into which to place weeds
Soil and water conservation measure to trap soil and conserve water	Barriers across the slope intercepting soil	Convenient way to subdivide garden for planting and management purposes
Danger of erosion and need to instigate organic conservation measures to decrease erodibility	Erodible soils	Opportunity to harvest sediment at bottom of slope and create new field

Scales of Field Assessment

Land degradation occurs at widely varying rates, and to varying degrees, over the landscape, hillside and between fields. As noted in the previous chapter, the focus of this handbook is on the local scale. Levels of degradation are considered by reference to farms and individual fields. In the case of rangelands, degradation refers to dispersed features, such as tree mounds and gullies. The local-scale focus, together with the farmer-perspective, dictate the type of measurements that are appropriate. The methods described in Chapter 4 of this handbook, in relation to the measurement of soil loss, are particularly suited to field and farm scale. They accommodate the fact that soil loss does not occur uniformly across plots or hillsides, and instead allow for the variability within the natural landscape that affects the amounts of soil loss and runoff from apparently homogeneous fields. They are also measurements of soil loss and accumulations that can be readily observed by the land user.

The perception of the scale and seriousness of land degradation will be influenced by the timing of any investigation. Many forms of soil loss are most easily seen during or shortly after periods of heavy rains. Some types of erosion may be less visible after crops become established in fields. Nutrient deficiencies and other factors that affect crop production will be best observed when crops are in-field and relative growth rates can be assessed. Actual production is best assessed at harvest times when output can either be weighed, or the standard number of units (sacks/bundles) counted. Repeated measurements give a more complete picture of the effects of the processes leading to land degradation.

The causes and effects of land degradation can occur both on- and off-site. On-site effects of land degradation lead to a lowering of the land's productive capacity, resulting in reduced yields or a need for higher inputs. These costs are borne directly by the land user, thus affecting interest in reducing or reversing land degradation. The land user's ability to remedy the land degradation depends on whether the cause is on- or off-site. Off-site effects of land degradation are problems exported and borne by others. The most common off-site effects include sedimentation in reservoirs and waterways, decline in water quality and contamination of drinking water, gully erosion and deposition of eroded materials on farmland.

Because the causes and effects of land degradation are unaffected by the boundaries of land ownership or use rights, degradation may occur on a farmer's land as a result of actions taken by other land users upslope. Similarly, actions taken on a farmer's field may affect other land users downslope. Therefore, the interest in preventing land degradation may not coincide with the cause. This has serious implications when it comes to assessing the costs and benefits of different courses of action. For example, if upstream soil erosion causes siltation of a reservoir, from the reservoir operator's point of view the net benefits that accrue from incurring expenses to reduce or eliminate the erosion may well outweigh the costs of doing nothing. However, the land user whose farm is the source of the deposited soil is unlikely to attach the same level of benefit to the reduction of soil loss.

Levels of Analysis of Degradation

The examination of field degradation at different scales feeds into different levels of analysis. Each level has its own particular set of uses. The first and most immediate use of information relating to existing or potential degradation is to identify the risks at field and farm level. Mapping of fields and detailed site inspection are involved here. The next level is to rank the degrees of actual degradation, or future risk of degradation, by reference to their seriousness. This allows the land user to prioritize possible responses to degradation risk and to target parts of the farm where risk is greatest. The field assessor may use this level of analysis to make semi-quantitative comparisons between sites and situations. A third level of analysis is to formalize the prioritization by farmers by attaching monetary values to the costs (time, labour, money) and to the benefits of any course of action (including 'doing nothing').

1 Mapping of fields

The first step in assessing land degradation is to take stock of the visual evidence of degradation in the area under review. The physical aspects of the landscape must be observed and evaluated. Preparing a map (see Figure 2.1) of the area under review (farmer's field or farm) will help to identify areas at particular risk of degradation due

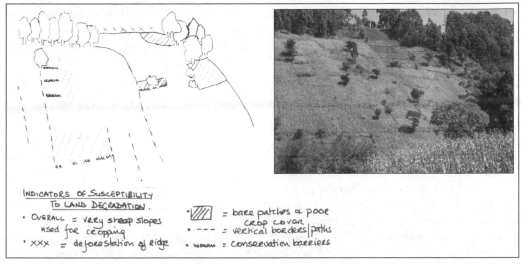

Figure 2.1 *Sketch of a Small-farm Agricultural Landscape in Kenya, Showing Susceptibility to Land Degradation*

to the naturally occurring features in the landscape. Discussions with farmers will furnish important information about yield and the vigour of plant growth in different areas of the field.

The site-specific characteristics identified at this stage help to show where the highest risks of land degradation lie within a field, farm or over a larger area. A systematic approach to mapping of the area under review will not only identify existing degradation but pinpoint areas at risk from future degradation. Since land degradation occurs as a result of the combined effects of soil characteristics, slope angle, climate and land management, changes introduced by the land user will affect the risks of land degradation.

The mapping of the area under investigation aims to identify the causes of degradation and to explain why some parts of the area under review may be more susceptible to degradation than others. Proportionally more effort may be required by the land user to protect susceptible areas from future degradation.

2 Ranking risks according to seriousness

Having mapped (in-field or on-farm) the actual degradation occurring and the potential for further degradation in the future, the identified risks can be ranked based on their seriousness. Chapter 6 gives some guidance on how this ranking can be carried out, not only to assess the risks but also to provide a tool to assist future decision-making. This ranking leads into action plans for combating land degradation, allowing land users to prioritize the focus of their conservation/land degradation prevention activities.

3 Cost-benefit analysis

The identification and ranking of the risks of land degradation form the data for further analysis. It enables farmers to estimate the costs and benefits of measures and techniques that will reduce or eliminate land degradation, and to compare these with the costs and benefits of doing nothing. Farmers typically choose between a number of options, including allowing land degradation to continue, undertaking low-cost but technically limited interventions or implementing a full package of high-cost measures. The field assessor should be able to provide the needed data for choosing which option best meets the farmer's needs.

This kind of assessment, known as cost-benefit analysis, underlies the process of making decisions about investment in land and farming activities in both smallholder and commercial agriculture. Whether or not to invest in a capital or labour intensive activity will depend on the perceived benefit of it to the person making the investment. This latter point is important – while economics enables us to carry out simulated cost-benefit analysis for decision-making purposes, ultimately the analysis is subjective, relying on the values attached to specific costs and benefits by individual land users. Consequently, two farmers living side by side, with similar farms in terms of area, topography and fertility, may make widely different decisions about land management issues, be it the crop to be planted, the fertility treatment to be undertaken or physical conservation measures to be dug. This subjectivity reflects the circumstances of the individual land user.

Cost-benefit analysis must not be seen as a prescriptive tool. It cannot be applied mechanically to arrive at a single 'right answer'. Capturing the costs and benefits that are important to the individual is the best way of getting close to the 'right answer' for that farmer.

This handbook will not deal with cost-benefit analysis in detail – it is really an extension of field assessment and a way of using data to gain a view of the likelihood of farmers' decisions on whether to invest. However, it is important that the field asses-

sor gains the information about the important variables for undertaking cost-benefit analysis, so that the analysis can be accomplished later using any one of the many manuals that describe how to do it. The variables of greatest importance for a farmer-perspective cost-benefit analysis are:

- **Costs:** these must reflect the real costs to the farmer of undertaking any protection measure against land degradation. The largest cost is usually labour, and the field assessor needs to get a good understanding of what other activities the farmer cannot undertake in order to accomplish the conservation (this is the opportunity cost of labour). Similarly, there are costs in land and capital, which must be realistically assessed. The input of farmers is vital in making these assessments.
- **Benefits:** these must also reflect the real benefits to farmers. There are direct benefits such as increased yields; but the indirect benefits can be larger. For example, reduction in weeding because of a good cover crop, or reduced ploughing costs because of better soil structure, are legitimate ways in which reduction in land degradation brings benefits to land users.

Other important variables include time horizon (what planning horizon does a farmer use?), the discount rate and the valuation approach. Guidance on these, and other issues relating to cost-benefit analysis, can be found in most economics textbooks. Several useful references are suggested in Appendix IV.

Cost-benefit analysis of land degradation is considered further in Chapter 8, in terms of appraising a conservation technology. The principles are identical, whether the assessor wants to know if land degradation is costly, or conservation is worthwhile.

What About the Land User?

First Consider the Land User

Although land degradation is a physical process, its underlying causes are firmly rooted in the socio-economic, political and cultural environment in which land users operate. For example, for some land users poverty may be a key factor that leads to land degradation since poor land users may become stuck in a cycle of degradation, where their poverty precludes investment in the land, lack of investment leads to further land degradation, and degradation to more poverty. Consequent upon the downward spiral are low crop yields, adverse food security and little surplus production for sale, thus reinforcing the poverty of the land user. Other issues such as security of tenure, alternative income-earning opportunities and labour constraints are additional land user factors important in determining overall land degradation status.

Farmer-perspective field assessment needs to recognize these complex relationships between the land and society, and how land users may find it economically rational to degrade their soils until conditions change that then induce them to protect against further land degradation. Only by understanding the forces influencing farmers' actions can the field researcher begin to comprehend the dynamic interactions between socio-economic factors and land degradation. With this realization, the researcher may start to appreciate the consequences of land degradation for land users (Chapter 7) and to address the design of interventions that bring benefits both to society and to land users (Chapter 8).

A classic example of how economic imperatives have conditioned people to degrade their land is found in southern Africa. Lesotho has the unenviable reputation of having the most severely degraded land on the continent (see Plate 9). This is partially explained by the underlying physical conditions (easily-erodible weathered basalt) and partly by poor standards of farming. Overgrazing by cows is endemic and conservation measures are routinely ignored despite substantial subsidies and campaigns by aid agencies and the government. Yet, Lesotho's human population is not particularly large and many of the soils could be quite productively farmed. So what is going on? Lesotho's so-called farmers are nothing of the sort – they are migrant labourers in South Africa, returning home for holidays, to bring up children and to live in retirement. Their money they bank in cattle to graze (and overgraze) the open access hills. For any individual it would be economic madness to devote time and resources to improving the land. The economic payback would be so small compared with the income of migrants working South Africa's gold reef or coal mines. To ignore this complex reality would mean a failure to appreciate *why* land degradation is occurring and *how* conservation measures would be spurned – that is, until the balance of economic investment changes to favour improving Lesotho's own land resources.

This example from Lesotho demonstrates that various factors can initiate and enforce land degradation. Land degradation has occurred, and continues to occur, in both developing and more developed countries regardless of political systems and wealth. An important distinction can be made since the proportion of the population directly affected by land degradation, to the extent that livelihoods are adversely affected or even threatened, is much greater in less developed countries than in developed areas. There is, therefore, a fundamental difference between a market-based developed economy and a subsistence-based developing economy. The first can weigh up the cost of ignoring degradation by paying to offset its effects against the value of production. In market economic terms it often makes sense not to address the degradation directly but compensate for it indirectly. The second has less choice – technological inputs are unavailable or too costly, so conserve or starve.

The field assessor needs to ask careful questions of local people, involving them diplomatically in the analysis of why land degradation may be happening. The Sustainable Rural Livelihoods (SRL)[1] framework is a useful platform for bringing the relevant issues together. The kinds of questions to which the field assessor will need answers are set out below. Clearly, this cannot be an exhaustive list since particular circumstances will warrant the collection of specific types of information. These questions are not designed to be asked directly of the land user, but are prompts to the assessor that the information is needed. The information must be collected in a way appropriate to the circumstances of the land user. Often a roundabout approach, involving a series of more simple questions, each building on the last, will be effective in eliciting information from the land user in a non-threatening way. (See Appendix IV for suggested reading.)

- What encourages you to protect your land from degradation? Income; value to your children's inheritance; pressure from other land users, the chief; subsidies to undertake conservation; inspection by the extension officer; pride and morality; and so on?
- What discourages you to protect? Economic opportunities elsewhere; poor market for crops; high cost of labour and/or implements for conservation; lack of land security; and so on?
- How is your livelihood supported by the natural environment? For example, local medicinal plants, good grazing resources, abundant fuelwood (or the opposites).
- How is your livelihood affected by your skills and knowledge? What about indigenous techniques of managing land resources, and your adaptations of recommended practices?
- How is your livelihood affected by the money you have available? Consider all sources of income, such as cash remittances, income from crops and livestock, selling of labour.
- How do other people locally help you? Relations, local societies, cooperatives? Do these enable you to carry out farming practices you could not do by yourself?
- How is your use of the land affected by other factors, such as markets, roads

and communications, availability of tools or advice, and ability to access the right seeds and information?

Considering these questions initially with local farmers will give the field assessor a much better grasp of what factors are important to the land user and how the presence or absence of these factors may induce or prevent land degradation. Such knowledge is just as important as direct measures of land degradation (Chapter 4) or its effect on production (Chapter 5).

Factors Affecting Land Users and Land Degradation

The following list gives an indication of the breadth of issues that affect land users' decisions about activities that may have a consequence for land degradation. Because they introduce factors which may control land users' priorities and practices, these issues are relevant, whether or not land users directly undertake conserving activities. For practical purposes, conservation is the reverse of degradation – the following issues may either encourage or discourage a farmer to undertake resource-conserving practices.

1 Land Tenure

Security of land tenure affects farmers' willingness to invest resources in land improvement and protection against degradation. Insecurity of land tenure  shortens the time-frame used by farmers for decision-making, making it less likely that measures which protect against land degradation will achieve a positive return in the planning horizon of the land user. Where the occupier of land is unsure of the future, extraction (or 'soil mining') will occur to ensure that these resources are not lost to the individual. A farmer with clear title to the land is more likely to consider investment of money, labour and land in

conservation because benefits in production which may only accrue after many years will still be retained by the individual who implemented the measures. Common property resources are especially vulnerable to land degradation. However, the field assessor needs to distinguish carefully between 'open access' where land users have virtually free rein to use whatever resources they can grab, and 'common pool' resources where access is controlled. Common pool resources are much the more prevalent, and local societies' means of controlling land-degrading activities on these resources should be assessed. A good example is the *ngitili* of northern Tanzania, which are dry-season grazing reserves held commonly by the local elders on behalf of the village community. All in the village have access to these, but this is carefully controlled to avoid the resource becoming overused, or one individual grabbing an excess share of the limited grazing.

2 Poverty

Poverty affects how land users manage their land. It reduces the options available, ruling out some conservative practices because they require too much investment of land, labour or capital. Similarly, poverty tends to encourage farmers to focus on immediate needs rather than on those whose benefits may materialize only in the long term. This is not to say that poor farmers are land degraders, while the  rich are conservers. Several studies have shown exactly the opposite. In Ethiopia, for instance, some poor farmers have been reported to invest more in their land than the rich, probably because they are almost wholly dependent on their land. The foreclosing of expensive land use options may make the poor develop and apply simple but very effective technologies such as trashlines, earth mounds and ridges, or intercrops. Poverty may also induce rural people to abandon farming and migrate to towns, with a consequent benefit to the land. What the poor cannot do is expend huge effort in digging bench terraces or hiring bulldozers. These measures, available only to the rich, may be effective in controlling land degradation, but they need continual maintenance and commitment by the land user – obligations which the rich may not be prepared to undertake – if they are not to fall into disrepair and induce further land degradation. Poverty is, therefore, a somewhat ambiguous factor, that needs careful analysis and interpretation in its effect on land degradation.

3 Pressure on the Land

A growing population puts greater demands on the land. Farms are split into ever-smaller units as land is shared out among family members. Land shortage acts as an incentive for land users to push the boundaries of cultivation into more marginal areas, less suited to continuous use. Increasing numbers of people require more food, more water,

more fuelwood and more construction materials, all of which must be sourced from the environment. An indirect effect of land pressure is the requirement for more extensive infrastructure. More roads, more transport, more housing and more utilities all have the potential to lead to increased land degradation. However, as with poverty, the evidence for a direct link between increasing populations and degradation is ambiguous. Indeed, several studies have shown how populations may adapt to new circumstances through developing new technologies and adjusting old ones. In some places, where markets and rural infrastructure have allowed, increased population density appears to have been the spur to sustainable intensification. Extensive land-degrading practices such as fuelwood extraction and large herds of livestock have given way to intensive, well-managed small farms, employing manuring, composting, agroforestry and other beneficial practices. So, care is needed before making specific judgements about the effects of population on land degradation – but the issue must still be addressed.

4 Labour Availability

Labour is normally the most limiting constraint of smallholder farmers. Competition for available labour is especially intense between laborious activities such as constructing terraces and off-farm employment that can bring immediate returns. The prevention of land degradation involves the investment of labour, both at the initial stages and on an ongoing basis for maintenance. Land users often overcome labour (and other capital) shortages by implementing conservation measures gradually, spreading the work over several seasons or years. Indirectly, the investment of family and hired labour is crucial to land degradation in enabling more intensive (and generally more conservative) production systems to be undertaken. Gender divisions of labour are also important: practices such as land preparation, tillage and weeding are normally assigned to one sex. For example, in Tanzania *ngoro* pits are traditionally constructed only by women (see Plate 10). If that gender has limited labour available at the right time, then there may be implications for land degradation which need to be noted.

In a situation where labour is already a scarce resource, it may not be possible to supply the additional labour required to avoid degrading activities or to undertake conservation. Migration to urban centres is a common feature of rural communities in developing countries. While this may reduce the immediate pressure in terms of the numbers to be supported from a single smallholding, the loss to labour may increase the risk of degradation.

5 Economic Incentives

There are a number of ways in which the markets may affect a land user's decision  about degrading or conserving farming practices.

- Price structures for agricultural produce often favour the urban purchaser over the rural vendor. As a result it may not be possible for a land user to recover the costs of more expensive non-degrading production methods in the selling price achieved for produce.
- Alternatively, quick profits may be possible by maximizing production in the short term. The effects of potentially degrading activities may be ignored or, where additional inputs such as fertilizers are used, masked.
- High risk may attach to agricultural production due to market volatility or political instability. Land users may be less prepared to invest in the land where the potential returns are uncertain.
- Economic instruments such as subsidies and other incentives distort farmers' priorities. Conservation measures in many countries attract direct financial inducements based on measurable values, such as metres of terrace or number of trees planted. Such distortions often carry through to the withdrawal of subsidies, when farmers are no longer prepared to practice conservation without payment – a situation that is now common in South Asia, leading to considerable worries about the effects on land degradation.

6 Indigenous Technical Knowledge

Through generations farmers have accumulated knowledge and experience based on  what they see that works well for their own situation. This indigenous technical knowledge (ITK) is now well recognized as a valuable commodity, tried, tested and adapted for local environments and accepted by local people. ITK methods such as the *Zaï* planting pits of Mali and Burkina Faso and the *kilimo cha vinyungu* or raised beds of the Wabera tribe of Tanzania provide forceful evidence that indigenous knowledge is especially suited to protecting against land degradation. With modernization and globalization the continuation of ITK is very much under threat.

7 Appropriateness of Introduced Technology

Technologies developed on research stations may prove to be inappropriate 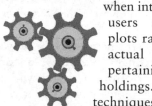 when introduced to land users since research plots rarely mirror the actual conditions pertaining to small-holdings. For example, techniques may take too much land out of production, need too much labour to construct or maintain them, or compete with crops for water or nutrients. Introduced techniques may be socially or culturally unacceptable as, for example, fenced paddocks for rotational grazing on common land (see Box 3.1). Where land users have had previous negative experiences with conservation technologies, they are understandably reluctant to adopt new conservation plans. Similarly, where previous conservation attempts have been ineffective either through poor design and inadequate extension or poor execution and maintenance, land users may be unwilling to invest time, effort and space in new technologies.

8 Economic and Financial Returns

Most decisions made by land users are based on economic rationality as perceived  by the land user. Such rationality controls the willingness to invest in any practice, especially in demanding measures needed for land degradation control. Where a farmer's individual cost-benefit assessment concludes that the benefits of a prevention/conservation course of action do not outweigh the costs, then the rational decision for that farmer is not to undertake the works. Where inse-

cure tenure is also a factor, the anticipated benefits are reduced by the short-term time horizon of the land user. Field assessment is usefully supplemented by relatively simple cost-benefit analysis techniques, such as discounted cash flow analysis. With farmer participation, the financial worth of investing labour, land or capital in any land improvement may easily be assessed, using criteria such as net present value, internal rate of return or returns to land/labour/capital. These techniques are beyond the scope of this handbook but the Annotated Bibliography (Appendix IV) and short discussion in Chapter 2 provide more guidance.

9 Off-site versus On-site Costs

Costs and benefits incurred on-site (the farmer's field, for instance) are private or personal to that land user. Costs incurred, say, as a result of sedimentation into dams and rivers off-site are a consideration for society. Few land users will be prepared to invest private resources solely for the benefit of society, unless society supports such activities through subsidies (see 'Economic Incentives' above). Where the land user does not bear the full costs of land degradation, the incentive to take action to reduce land degradation may be insufficient for the land user to change practices or adopt new technologies. Costs that are incurred downstream of a land user's plot are unlikely to be incorporated in land use decisions. The field assessor needs to note where the land user's activities are having an effect – on-site or off-site – and who is being affected.

10 Power and Social Status

Some components of production are driven by a need to preserve social stand-

BOX 3.1 RANGELAND CARRYING CAPACITY: SCIENCE VERSUS SOCIETY

Land carrying capacity is a straightforward scientific concept devised by rangeland scientists anxious not to allow the land to degrade. Grass has a finite production of biomass. Exceed the off-take of forage by animals through overstocking and the vegetation (and eventually the soil) degrades. Typically, rangeland carrying capacity in the drylands of Botswana is one livestock unit (equal to one cow) per 10 to 20 hectares.

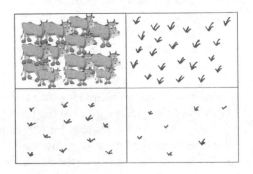

To stock the range at its optimum, rangeland scientists in southern Africa devised rotational grazing schemes – known as the *Savory* system – where cattle graze out one paddock, then are moved to another and then another. Palatable and unpalatable grasses alike are eaten. Each paddock has sufficient rest period to restore both the quantity and quality of its vegetation.

Nice idea! So why has it not been accepted by pastoralist communities on extensive tracts of common land? First, fencing is anathema to common property access and use. Posts and wires are the instruments of land alienation. In one of Zimbabwe's communal areas fencing wire was stolen within a year – despite community agreement and participation in the scheme.

Second, the formal science contradicts indigenous reasoning and technical knowledge. There is no fixed carrying capacity – both the biophysical and climatic environments are far too variable in space and time. Herders have learnt to be opportunistic; that is, in times and places of plenty, to have as many animals as they can get. Not only does this add to the individual herder's prestige in society, but it is also rational. Come hard times such as drought, many animals die – but a few lean ones survive. There is a far greater chance of some survival if you have many animals than if you have few. Additionally, there is no substantive evidence that such an opportunistic strategy is any worse for rangeland productivity – if anything, the opposite may be true, controlling bush encroachment and keeping out tsetse fly.

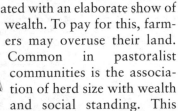

ing or to enhance prestige. In some cultures weddings and funerals are associated with an elaborate show of wealth. To pay for this, farmers may overuse their land. Common in pastoralist communities is the association of herd size with wealth and social standing. This association is one of the reasons why herders deliberately keep as many animals as possible, despite their impact on rangelands. The field assessor needs to be aware of cultural traditions in so far as they affect land use decisions (Box 3.1).

These factors are not mutually exclusive. They may be cumulative and interactive. They all need attention as part of the diagnosis of why and how land degradation is occurring or not occurring.

Sustainable Rural Livelihoods (SRL)

Categories of Asset

In looking at land degradation, the purpose is not only to determine whether land degradation has been, or is, occurring. Any consideration of land degradation must also address the root causes of the degradation and ultimately seek ways in which the degrading activities can be reversed. Many rural livelihoods depend on the natural environment, thus any permanent diminution in the productivity of that environment will have adverse effects on the ability of families/household units and communities to support themselves.

The factors that affect the decision to degrade or conserve land are related to the resources available to the land user. Resources have been subdivided in what is known as the Sustainable Rural Livelihoods framework into a number of different elements or 'capital assets'. These categories of asset can be used to describe the various types of 'capital', or resources, available to land users. As such, they provide a framework for analysing the situation of land users, which may be helpful in identifying sets of circumstances that combine to make some households more likely to degrade their land than others. The diagram in Figure 3.1 summarizes the five categories of capital.

In general, individuals, households and communities have different access to each type of capital. Lack of one category of capital may be compensated for by another. For example, physical capital in the form of a plough may take the place of human capital where there is a shortage of labour. One form of capital can be converted to another. Financial capital (cash) may be used to acquire human capital (labour), physical capital (fertilizer) or natural capital (land or livestock). Similarly, social capital, through group membership, may make it possible to draw on community labour at harvest or other busy times.

Access to capital assets is prescribed by society, by way of formal rules and socio-cultural norms. Thus factors such as gender relations, marital status, education,

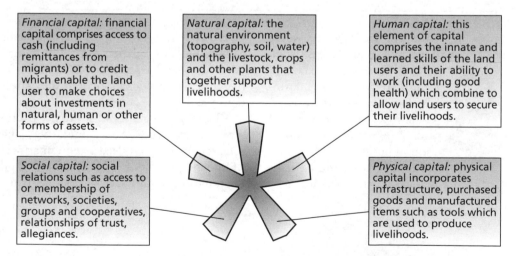

Financial capital: financial capital comprises access to cash (including remittances from migrants) or to credit which enable the land user to make choices about investments in natural, human or other forms of assets.

Natural capital: the natural environment (topography, soil, water) and the livestock, crops and other plants that together support livelihoods.

Human capital: this element of capital comprises the innate and learned skills of the land users and their ability to work (including good health) which combine to allow land users to secure their livelihoods.

Social capital: social relations such as access to or membership of networks, societies, groups and cooperatives, relationships of trust, allegiances.

Physical capital: physical capital incorporates infrastructure, purchased goods and manufactured items such as tools which are used to produce livelihoods.

Figure 3.1 *The Sustainable Rural Livelihoods Framework*

caste and age influence access to assets and services. Within a household, too, access to assets is an ever-changing equation, determined both by social conditioning and by relations between household members. Levels of capital assets are not static but change from season to season, and from year to year, as a result of actions both by household members and by agents outside the household, at community, regional or national level.

The concepts of sensitivity and resilience, discussed in connection with the effects of changes and shocks on landscapes and ecosystems, can be applied, in much the same way, to the livelihoods of individuals, households and communities. These livelihoods are also more or less sensitive, and more or less resilient, to changes or shocks. The sensitivity and resilience of households may be directly related to how they deal with the capital resources available to them. Shortages of one or more types of capital may increase the risks of shocks and changes.

The SRL Framework and Field Assessment of Land Degradation

The SRL Framework gives a useful means of organizing the many types of information relating to the land user, the production system and their potential influence on land degradation. In particular, the framework can highlight circumstances that make land degradation one possible outcome of future activities, or where a transfer of capital from one type to another may affect the potential for degradation. The intention for the field assessor of land degradation is *not* to undertake a full livelihoods analysis, which is beyond the scope of this publication, but to present a systematic coverage of the aspects of the land user's circumstances that control the biophysical processes of land degradation. The objec-

tive is to collect data potentially useful to support the more direct field assessment methods in Chapters 4 and 5, and to provide the explanation for the patterns of land degradation observed.

Table 3.1 illustrates how land degradation could be considered in conjunction with the capital assets framework. It gives examples of how changes in the level of assets available to a household may affect both land degradation and conservation. It is important to note that increases in a particular capital asset do not automatically lead to less land degradation or more conservation. Nor does the converse hold true. There are many other factors, not least the interaction with other capital assets, which affect the outcome of changes in capital assets.

Table 3.1 enables the field assessor to note various positive and negative elements of capital assets in relation to their potential influence on land degradation. The SRL framework should also encourage comparisons between the situations of different land users, and over time. In some cases, the relative capital wealth of a household will be evident. For example, the comparison of a landlord with a landless peasant may indicate that the former has a greater capital stock than the latter. However, because capital can be added to, or lost, the balance between these two individuals may change. If the landlord's position in the community were undermined (social capital), for instance by a change in government, this would equate to a depletion in the landlord's capital wealth. This might have knock-on effects on the willingness of labourers to work for him (thus affecting human capital) which in turn could necessitate the payment of higher wages to those labourers, reducing the landlord's financial capital. Conversely, a landless peasant may substitute his or her human capital for natural capital. Thus, the peasant's livelihood may

Table 3.1 *Illustration of the Field Assessment of Capital Assets*

Capital Asset	Examples of How Land Degradation and Conservation Might Be Affected By:	
	Increasing Capital	Decreasing Capital
Natural	Extensification of farming onto larger areas of land leads to poorer land management and more degradation Increased livestock numbers contribute to additional land degradation More conservation undertaken if land is no longer a limiting factor	Intensification onto smaller units of land results in more conservative practices and less degradation Greater production required off a smaller portion of land, leading to 'soil mining' and degradation Reduced levels of livestock lead to less land degradation Greater efforts may be made to conserve the remaining natural asset base
Physical	Labour-saving farming practices may enable more time to be spent on conservation Inappropriate technologies may increase the risk of land degradation	Deteriorating local roads lead to reduced market opportunities and lack of investment in land management Lack of spare parts for tractors means no maintenance of conservation structures, breakage in storms, and severe degradation Lack of spare parts for tractors leads to a return to animal traction, reducing compaction and decreasing degradation
Human	New techniques and skills may be applied to land management practices, resulting in less degradation and/or more conservation New skills or education enable family members to take off-farm employment, reducing the labour available to undertake farm work and increasing degradation New skills in farming enable better practice and reduced land degradation	Out-migration reduces demand from the land, leading to less land degradation Out-migration reduces labour availability, leading to poor farming, more degradation and less conservation AIDS/HIV kills active farm labour, causes land abandonment and decreases land degradation
Social	Admission to a cooperative may provide access to better information, technologies or community labour to take action against land degradation Marriage may strengthen kin networks and foster new relationships and allegiances which may be called upon to supplement family labour for the construction of conservation works Involvement in local politics increases the time spent away from the farm, reducing labour availability for construction and maintenance of conservation works, resulting in more land degradation	Disputes with neighbours may isolate a household and make it difficult to access community labour groups, for example to undertake planting, harvesting or conservation works Divorce may affect the ability to draw on kin networks at times of stress Disputes with neighbours may also reduce the time spent on communal works, allowing the farmer to concentrate on better land management practices

Financial	Increased access to finance/credit enables land users to undertake expensive conservation works Increased access to finance/credit enables land users to buy large 4WD tractor to plough much more extensively and deeply in all weathers, thereby increasing land degradation Increased remittances from urban-based family members allow farmers to divert attention from the land and encourage poor standards of farming	Sudden decrease in income results in plundering of natural assets or the diversion of essential labour to meet essential expenditure Reduced availability of credit for fertilizers forces farmers to rely on compost and manures, thereby reducing land degradation Reduced availability of credit for fertilizers leads to poorer crops, more erosion and increasing land degradation

be secured, and financial capital accumulated, through the use of skills and knowledge in paid employment.

Because capital is continually changing over time, and because there are so many different components of each type of capital, initial observations concerning access to capital may be misleading. In addition, how these different types and components of capital can be combined is a difficult question. Ultimately this is dependent on the importance to livelihoods of particular components of capital, in specific circumstances. Table 3.1 illustrates several examples where the same change in a capital asset could have a positive effect on land degradation for one person but a negative effect for another. The analysis of the whole picture requires a detailed understanding of people's livelihoods.

For the field worker, Participatory Rural Appraisal (PRA)/Participatory Learning and Action (PLA) can provide insights into local people's perceptions of their circumstances and the possibilities open to them. (PRA is about the appraiser learning from the farmer, while PLA has a greater emphasis on shared understanding being turned towards improvement of the farmer's situation. Both are involved in this handbook, but collectively their techniques are referred to as PRA tools.) Some of the techniques may be useful in identifying capital assets. Thus, PRA tools may be used to discover factors of the land user which impinge on decisions which might alter the status of land degradation. PRA may result in local people, themselves, determining how best to deal with land degradation problems and how to select between possible conservation solutions.

Participatory Land Degradation Assessment

PRA is 'a family of approaches and methods to enable rural people to share, enhance and analyse their knowledge of life and conditions, to plan and to act' (Chambers, 1994, see Appendix IV). PRA tools typically involve local people in the identification of an issue, such as land degradation, the assessment of its impact on their livelihoods and the selection of the most appropriate means of addressing the problem identified. The participatory approach seeks to involve all groups in society – men and women, young and old, rich and poor. Different perceptions by different groups of people can then be taken into account in selecting the most appropriate solutions. Plate 11 illustrates the importance of discussions with people in the field.

PRA tools can be divided into several categories depending on their purpose. Some are designed to discover the ways in which rural people perceive and use **space** and **time**. There are other tools for establishing preferences and differences (**ranking and classification**), for describing and understanding linkages (**flow diagrams**) and for establishing **decision-making** processes. The attributes of each category of tool are:[2]

Space

Rural people often allocate space in intricate ways, especially if there is strong differentiation in quality of land and access to it. PRA tools such as sketch maps and transects can be used to compile an inventory of resources. The objective of, for example, a sketch map derived in a participatory way is to arrive at rural people's perception of their natural resource situation. Maps and transects can provide complex information such as who uses a particular resource, when and how. The advantage of maps is that they break down communication barriers, help to focus attention on issues to be discussed later, encourage observation of things which are not normally even thought about by local people or the field assessor, and ensure that diversity is taken fully into account. Transects are in effect systematic walks through an area, even in the rain (see Plate 12), to note community land use and practices and to compile detailed spatial information. Their principal advantage is that 'outsiders' such as the field assessor are aware of all village land use activities as a baseline for further enquiry.

Time

'Time tools' are probably the best known, most used and most diverse. They include calendars, historical profiles and timelines. They are used to record change over time of many events such as pests and diseases, food availability, progress of a gully, deforestation, as well as changes in important explanatory variables such as population growth and droughts. The advantage of time tools is that a range of information on a number of issues can be gathered in a relatively short time. They can be used to test possible relationships, such as change in agricultural practices and soil quality. The value of the information depends on memory recall, and the further back in time, the poorer the recall. Nevertheless, the validity of time tools is enhanced in that they are usually used with groups of people rather than individuals.

Ranking and Classification

The use of ranking tools generally has been described as 'playing analytical games'. The simplest is the ranking of problems or attributes as first, second, third order of importance/seriousness. More complex tools allow for more description or exceptions. Ranking tools are especially useful for monitoring and evaluation exercises where, say, the applicability of a set of soil conservation technologies is being discussed. Stakeholder analysis is a particular form of classification tool that enables identification of all interest groups, with the view to aligning interventions to the needs of individuals or groups in society. In use in PRA, the classification of stakeholders is best done to identify the major differences in perception, attitude, resources and capabilities of various groups recognized by local people. Wealth ranking is one type, relating to capital assets and how people view others in relation to themselves. This sort of ranking and classification has been used to stratify subsequent samples of the population to ensure participation by all groups in the community, to establish the criteria which a community uses to differentiate its own population, and to establish who gains and who loses by any activity.

Flow Diagrams

Establishing links between activities, events and outcomes is an essential part of rural analysis. Tools for this include systems diagrams, problem trees and simple activity flow charts. The first is best known, where for example links between parts of a farm and aspects of the household livelihood can be established. Flow diagrams are an efficient way of identifying links where problems may be occurring, such as illegal cultivation of steep slopes because of lack of land. Problem tree analysis can be used to gather possible causes of problems and to guide the investigator to possible options for intervention.

Decision-making

This category of tools is used to describe sources of decision-making within communities and decision-taking steps for particular activities. Venn diagrams describe the decision-making groups, and the relationship between these and other groups. Decision trees describe the implications of specific decisions on resource management, and are useful for ensuring that all decisions required to achieve an outcome are taken into account.

In using PRA tools and techniques, information is obtained using semi-structured interviews, interviews with key informants and group discussions. In Plates 11 to 14, field professionals are shown using differ-ent interview techniques when meeting with farmers. Generally, it is best to meet with the land user in the field rather than in the home, and it can be instructive to go to the field when people are working there and when it might be raining.

- **Semi-structured interviews:** interviews with land users are important to gain an understanding of individual motivations and the rationale for particular courses of action or inaction.
- **Interviews with key informants:** discussions with community members can yield important insights into the social and economic structure of the community (Plate 13). Local names for soils and plants can be identified, along with key aspects of how features of land degradation have changed over time.
- **Group discussions:** Social groupings and how they affect access to and control of assets can be identified through group discussions (Plate 14).

These categories of tool, and the specific tools themselves, aid the field assessor to gain a far better understanding of important factors related to land degradation, especially attributes of the community and how various stakeholders perceive their situation in relation to the quality of the land.

Table 3.2 illustrates how a number of different PRA tools and techniques could be used to investigate the different types of capital asset described in the SRL framework. The table looks at how these tools can be applied at the household and at the community level. The scenarios in this table are examples only and are not an exhaustive list of how to apply PRA techniques to investigate capital assets – each situation requires its own careful, structured analysis.

PRA tools can be used to allocate land users into groups with similar attitudes,

Table 3.2 *Investigation into Capital Assets Using PRA Tools*

Tool	Capital Assets	Investigation Household	Community
Mapping	Natural	Farm layout, access to water, roads	Land uses, water sources, common property
	Social	Relationships between household members	Kin-based networks, other social groupings
Timelines	Natural	Changes in farm size, adoption of new crops or cropping practices, fertilization techniques	Changes in productivity, soil quality, climate
	Financial	Credit or cash remittances available on time for planting	Changes in access to banks, savings, credit institutions
	Social	Marriages, deaths, number of dependants	Cooperative networks, local institutions
Wealth Ranking	All	n/a	Local people's perceptions of relative resource endowments
Ranking and Scoring	Physical	Importance of access to tools/ farm machinery	Access to, and cost of, infrastructure
	Financial	Importance of different cash crops	Relative importance of different sources of cash and credit
	Natural	On-farm variety of crops, trees and other useful species	Species diversity and abundance

---

BOX 3.2 DISCOVERING PARTICIPATORY TECHNIQUES

During the last of the testing and validation workshops for the development of this handbook, a working group of Ugandan professionals were questioning Mr Nzirwehi, a Bushwere farmer, about the yield of bananas in different parts of his field where different degrees of land degradation were being observed (see Chapter 5 on production constraints).

However, there were only a few fruiting stems in the field, none of which could be accessed. So with the farmer, our Ugandan colleagues developed a new technique. With his hands the farmer indicated the size of banana bunches and the size of individual bananas in relation to position in the field. He also showed the professionals how the banana pseudostem was thinner in relation to bunch size. With these measurements and knowledge of the weight of a banana, the professionals developed a calibration curve of pseudostem circumference vs. yield of bananas. Armed with this, they conducted a yield assessment in relation to land degradation via the simple and accessible technique of measuring pseudostem circumference.

---

approaches and resources. They may identify risk areas, not only in terms of land characteristics, but also social, cultural and political circumstances within the local community. Appendix IV includes suggested further reading on these and other PRA techniques. However, the best way of learning participatory land degradation assessment is to do it (Box 3.2). To discover the many surprising insights revealed through participatory interactions is the best introduction to making more accurate land degradation assessment – the subject of the next chapter.

4

Indicators of Soil Loss

Land degradation encompasses a vast array of biophysical and socio-economic processes, which make its assessment difficult to encapsulate in a few simple measures. It occurs over a variety of timescales – from a single storm to many decades. It happens over many spatial scales – from the site of impact of a single raindrop through to whole fields and catchments. Without extreme care, measurements undertaken at one set of scales cannot be compared with measurements at another. This is why this handbook:

- adopts a farmer-perspective and hence the type of assessments at field and farm level, and timescales that have significance for the farmer (though, of course, these also vary according to individual circumstances – for example, the impact of erosion on the current crop, or concerns for long-term sustainability);
- focuses on the concerns of land users – primarily the way that land degrada-

tion makes farming more difficult, and the impact of degradation on productivity;
- concentrates on relatively simple field indicators, some of which can be quantified into absolute rates of soil loss, but none of which should be taken in isolation. The indicators are ones that farmers have told us they notice, and therefore information on them is more readily available. The indicators are also readily discerned in the field, although they are more apparent at some times of year and in some environmental circumstances.

This chapter addresses this last point – the development of field indicators of soil loss. The following pages have been organized to present a comprehensive set of 16 indicators: what they are, what processes lead to their biophysical formation in the field and, crucially, how to measure and interpret their meaning for land degradation. It will be seen that they apply to different timescales: an armour layer forms after

---

BOX 4.1 INDICATORS OF SOIL LOSS FROM RANGELANDS

Much of what is described in this chapter applies principally to cropland. Rangeland degradation, however, has been a major concern of scientists and land planners for at least two reasons. First, it tends to happen in dry areas where the ecology is extremely fragile and degradation rates probably at their highest. So it is a real process that is clear to see. Second, domestic animals, especially goats, have often been condemned as degraders of drylands. There have been many efforts at trying to destock to reduce rangeland degradation – usually with little success.

The principles, and some of the specific techniques, described in this chapter can all be used on rangeland. Vegetation, especially bare patches, and evidence of wind erosion are further indicators that can easily be developed for rangeland conditions.

Table 4.1 *Typical Measures of Bulk Density*

Soil	Average bulk density (t/m³)	Typical range of bulk densities (t/m³)
Recently cultivated	1.1	0.9–1.2
Surface mineral soils – not recently cultivated and not compacted	1.3	1.1–1.4
Compacted		
sands & loams	1.7	1.6–1.8
silts	1.5	1.4–1.6
clays	1.3	very variable

Source: Booker Tropical Soil Manual
Note: For the purposes of this handbook an average bulk density of 1.3 t/m³ has been used

one or two heavy storms in the early growing season, while tree mounds may take 50 or more years to form. They apply to different spatial scales: a soil pedestal may be only 1mm high, whereas a gully can, in exceptional cases, be 5km long. It should be clear, then, that none of these measures is directly comparable with another. However, after careful scrutiny, they can be used to ascertain general trends in land degradation. Tree mounds formed under trees of different ages can tell us whether degradation is getting worse, staying more or less the same, or even starting to reverse. Rock exposures and solution notches can also yield information useful over longer timescales. In contrast, the build-up of soil against field barriers such as boundary walls tells us what has happened in that field since the walls were constructed.

The following pages cannot accommodate all the observations that could potentially be made. In addition, they do not cover all the possible sources of error. For example, the value used for bulk density can affect the calculations of soil loss (see Table 4.1). Where the field assessor has been alerted to possible measurement errors, common sense will find many more and mitigation measures can be designed to minimize the unreliability. Experience has suggested that where one indicator is identified, the keen observer then starts to see many more. On the PLEC project in Yunnan, China, a short training course with local professionals found 12 different indicators in one field of 3 hectares in less than an hour. Our collaborators told us they had never before seen these features, and now wonder how they overlooked them! We advise that field assessors should start by using this chapter as a checklist of possible pieces of evidence of land degradation – we are sure they will find many more.

Participatory Field Sketch

The first and possibly most important step is to gain an overview of the field situation with the farmer. Experience has shown that drawing a field sketch map with the farmer has many advantages that can be used later when detailed assessments are made:

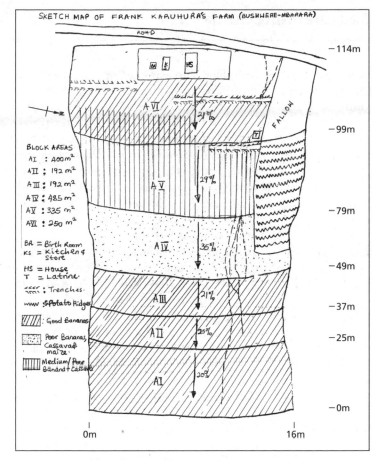

Figure 4.1 *Sketch Map of Mr Karuhura's Farm, Bushwere, Uganda*

each field type. The farmer will already have given an overview of the problems and potentials of the unit.

Sketch maps need only give an approximate idea of relative areas – see the example from Uganda (Figure 4.1). The most important aspects to include are:

- the spatial arrangement of field types with the homestead, road and other basic features;
- observations on key biophysical aspects such as slope, soil, rocks, springs – but only where these are important to land degradation and conservation;
- an initial record of the farmer's assessment of limitations of each field type and how land use has changed.

Armed with the sketch map, the field assessor is then able to plan a transect or sample design to look in detail at soil loss indicators in the key field types. The exact locations from which sample measurements are taken should be marked on the sketch map, allowing the field assessor to keep a record of these sites. This acts as an aide-mémoire when away from the farm and allows for checking of measurements that appear inconsistent. Additionally, these measurement points can be used again if the field assessor decides to re-visit the farm after some time to assess the effectiveness of new land management techniques or conservation structures.

- It is a good 'ice-breaker' – if the farmer is positively engaged with an output from the field visit which he or she can keep afterwards, then that alone brings the farmer onto the professional's side in a shared experience.
- It immediately identifies the 'field types' or the land units recognized by the farmer as being distinctly different by virtue of their biophysical condition (slope, soil) or their actual or planned land use (home garden, perennial crop plot, grazing paddock, etc).
- It provides a baseline from which to organize the detailed assessments for

Measurement Guidelines

In the field it is easy to become confused with the various scales of measurement. Here are some useful guidelines and conversions.

- **Small length measurements:** (less than 1 metre) for pedestals, rill cross-sections and similar features, use millimetres, mm
- **Large length measurements:** (over 1 metre) use metres, m
- **Area measurements in the field:** use square metres, m^2
- **Area measurements in soil loss results:** use hectares, ha
- **Mass measurements in the field:** use kilograms, kg
- **Mass measurements in soil loss results:** use metric tonnes, t (NB in other books, tonnes are sometimes written as Mg, ie a million grams)
- **Soil bulk density:** (needed for conversion of soil volume to mass) use tonnes per cubic metre, t/m^3 (In the example field forms, an average bulk density of $1.3t/m^3$ has been used – but see Table 4.1)

Sampling is an aspect of measurement that should be given close attention. For most indicators this handbook recommends a minimum of 20 observations. Sometimes, as for example with tree root exposure, it may not be possible to find 20 cases in the field. There is a trade-off between size of sample and validity of results. Because these indicators are designed to be quick to measure, the recommendation is always to do as many as possible.

A further measurement issue relates to the implied accuracy in citation of results. It must always be remembered that field measurements are 'rough and ready'. Therefore, in the rill example used later in this chapter the soil loss in m^3/m^2 is calculated as 0.000525. Converting to mass units, this translates to 6.825t/ha to three decimal places. But think what is behind such a calculation: for example, an assumed soil bulk density of $1.3t/m^3$. If that bulk density were out by $\pm0.1t/m^3$, the calculated soil loss would be 6.300–7.350t/ha. Therefore, it is quite erroneous and misleading to quote results to an implied accuracy such as this. At best the calculated 6.825t/ha should be cited as 6.8t/ha, but possibly even as 7t/ha.

BOX 4.2 USEFUL CONVERSIONS

mm ➡ metres	1mm = 0.001m (1000mm = 1m)
m^2 ➡ hectares	$1m^2 = 0.0001ha$ (10,000m^2 = 1ha)
mm lowering of ground surface ➡ equivalent in t/ha	1mm = 13t/ha (assuming a bulk density of $1.3t/m^3$)
Soil loss m^3/m^2 ➡ soil loss t/ha	$1m^3/m^2$ = 13,000t/ha (assuming a bulk density of $1.3t/m^3$)
π	$^{22}/_7$ = 3.14 (needed for calculations – see field forms)

Rills

What are they?

A rill is a shallow linear depression or channel in soil that carries water after recent rainfall. Plate 15 shows an example from Lesotho. Rills are usually aligned perpendicular to the slope and occur in a series of parallel rill lines (see Figure 4.3).

How do they occur?

A rill is caused by the action of water. Runoff is channelled into depressions which deepen over time to form rills. A rill is, then, a product of the scouring action of water in a channel. It is also a means of rapidly draining a small part of a field and efficiently transporting sheet eroded sediment from the rill's catchment. A broadly accepted distinction between rills and gullies, often applied in soil conservation, is that the former can be eliminated using normal agronomic practices (such as ploughing), whereas gullies require specific large interventions such as bulldozers, concrete lining or gabions (rock-filled bolsters placed in gully to accumulate sediment). Rills tend to occur on slopes, while gullies occur along drainage lines.

Where do they occur?

Rills will occur on a sloping surface where runoff is prevalent because of land use and lack of vegetation. Typically, rills occur where soil has been disturbed but the surface is left relatively smooth and unvegetated (eg after tillage, after building construction and on the sides of earth dams and road embankments). Rills are also likely to form in any slight depression in the soil, so paths, roadways, culverts and tracks made by tillage equipment are at risk of developing into rills.

How can they be measured?

The most common assessment of rills is the volume of soil that has been directly eroded to create the rill: ie the space and associated mass of soil now missing because of the rill. This calculation does not include any estimate of the amount of erosion that occurs between rills, ie inter-rill erosion, which can be measured using other techniques such as pedestals. The measurement of soil loss from rills assumes that the depression forms a regular geometric shape. Triangular (see Figure 4.2), semi-circular and rectangular cross-sections are most common.

In order to calculate the quantity of soil lost, it is necessary to measure the depth, width and length of the rill. A number of measurements of both the width and depth of a rill are suggested in order to get an average cross-sectional area. This averaging is appropriate as a rill will not be a constant width or depth throughout its length. These measurements of average cross-sectional area and length are used to calculate the volume of soil displaced from the rill. If it is known how long it has taken for the rill to form (if, for example the land was last tilled two months or two years ago), then an annualized rate of soil loss can be estimated.

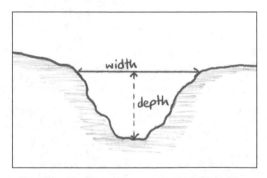

Figure 4.2 *Sketch Showing Cross-section of a Triangular-shaped Rill*

Single rills are rarely found. They usually occur together in the same part of the landscape. Each rill has a contributing area from which water will run off and pass into the rill, and sediment will be derived that is similarly passed along the rill. The most useful measure of the degree of importance of rill erosion is to calculate the volume or mass of soil per square metre of catchment (see Figure 4.3 and Box 4.3). This can be converted to tonnes per hectare to make the measurement comparable to other estimates of soil erosion.

Potential for error

1 Where rill erosion is evident, this is not the only form of erosion occurring. Rills are merely a visible symptom of sheet erosion. Therefore, it is important that any measurement of soil loss from a rill should not be treated as the total amount of soil lost from a particular area. The rill is indicative of the poor state of the immediate catchment of the rill and, wherever feasible, field assessments of sheet soil loss should be made. Experience indicates that the soil removed to form the rill is usually only a small fraction of the total soil loss from the catchment of the rill. This may not be the case if there is a dense network of rills.

2 Averaging cross-sections down the length of the rill, and then multiplying by the length of the rill, will give only an approximation of total volume. The more measured cross-sections and the closer the measurements are to the actual shape of the rill, the more accurate will be the rill erosion estimate.

3 As noted, rills occur where pre-existing depressions have become eroded by flowing water. The field assessor needs to estimate the volume of the original depression, and subtract this from the total volume, to calculate the soil removed by the rilling process.

4 Where re-deposition of the materials removed from the rills occurs in the same field, to avoid overstating the level of soil lost an estimate of the amount of soil re-deposited must be subtracted from the calculated soil loss from rills.

5 Rills are ephemeral features, easily obliterated by farming practice such as weeding. The evidence of erosion can, therefore, also disappear unless rapid and timely assessments are made. The early growing season in arable crops is especially conducive to rilling.

6 Estimation of the contributing catchment area to a rill must be made only after careful site inspection (see Box 4.3). Examine evidence of flow lines of

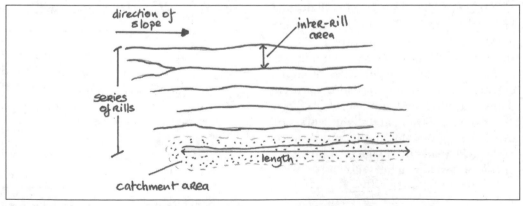

Figure 4.3 *Sketch: Series of Parallel Rills*

BOX 4.3 ESTIMATION OF CONTRIBUTING AREA

Rills act like mini-drainage basins, so they have a catchment of contributing area. This handbook suggests that field assessment examines the contributing area and expresses the soil loss by rilling initially as cubic metres of soil lost per square metre of contributing area. This, then, is converted to the equivalent in tonnes per hectare – see Field Form.

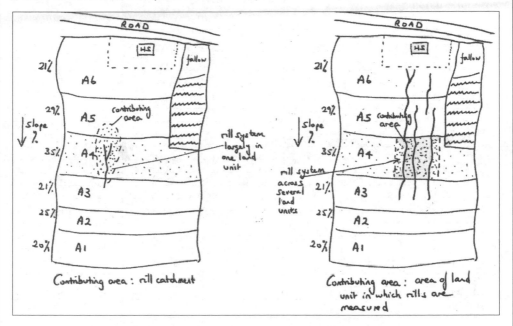

Figure 4.4 *Contributing Area of Rill Systems*

However, this does not always work where farms have complex patterns of field types and where the effective contributing area may be in one or more upstream field units or even on another farm. In such cases a modification is suggested; rill erosion should be expressed as cubic metres of soil lost per square metre of **field type** or **land unit**. In the left-hand sketch of a Ugandan farm, the rill system is largely contained within land unit A4 and the contributing area is the shaded area from which water runs off into the rill. In the sketch on the right, the rill system cuts across several land units. In this case, take only the area within each land unit affected by rilling as the contributing area.

This modification entails exactly the same measurement of rill volume. But the area over which that loss is assessed now becomes the immediate area of rilling. The distinctions should be clearly shown on the Field Form.

Source: Land Degradation Field Assessment Workshop, Evaluation report, Uganda, February 2001

water to determine the shape and size of the boundary of the contributing area. Look for the watershed between two rills as the boundary lines between contributing areas. In a levelled field between terraces or field edges, this is not usually difficult. The contributing

area may be of the order of 10 to 100m².

7   Rills may be caused (at least in part) by run-on from areas upslope. This should be taken into account when surveying for contributing area.

Worked Example

Field Form: Rill

Site: Farm of Mr Kashangirwe, Bushwere, Uganda
Date: 13 February 2001

Measurement number	Width mm	Depth mm
1	92	55
2	136	52
3	124	38
4	117	47
5	103	52
6	112	54
7	128	48
8	109	32
9	130	33
10	121	42
11	113	31
12	117	45
13	126	36
14	133	25
15	144	41
16	138	43
17	109	37
18	112	54
19	121	33
20	115	42
Sum of all measurements	2,400	840
Average (mm)	WIDTH = 120	DEPTH = 42

Length of rill: (m)   2.5
Contributing (catchment) area to rill: (m²)   12.0

Calculations

1   Convert the average width and depth of the rill to metres (by multiplying by 0.001). Thus, an average horizontal width of 120mm is equal to 0.12m and an average depth of 42mm is equivalent to 0.042m.

2   Calculate the average cross-sectional area of the rill, using the formula for the appropriate cross-section: the formula for the area of a triangle ($\frac{1}{2}$ horizontal width x depth); semi-circle ($\frac{1}{2}\pi$ x width x depth); and rectangle (width x depth). Thus, assuming a triangular cross-section, it is:

$\boxed{\frac{1}{2}}$ x WIDTH (m) $\boxed{0.12}$ x DEPTH (m) $\boxed{0.042}$ = CROSS-SEC AREA (m²) $\boxed{0.00252}$

3   Calculate the volume of soil lost from the rill, assuming that the measurements above were taken from a rill measuring 2.5m in length.

CROSS-SEC AREA (m²) $\boxed{0.00252}$ x LENGTH (m) $\boxed{2.5}$ = VOLUME LOST (m³) $\boxed{0.0063}$

4   Convert the total volume lost to a volume per square metre of catchment.

VOLUME LOST (m³) $\boxed{0.0063}$ ÷ CATCHMENT AREA (m²) $\boxed{12}$ = SOIL LOSS (m³/m²) $\boxed{0.000525}$

5   Convert the volume per square metre to tonnes per hectare.

SOIL LOSS (m³/m²) $\boxed{0.000525}$ x CONVERSION TO t/ha $\boxed{13,000}$ = SOIL LOSS (t/ha) $\boxed{7}$

Gully

What is it?

A gully is a deep depression, channel or ravine in a landscape, looking like a recent and very active extension to natural drainage channels. Gullies may be continuous or discontinuous; the latter occurs where the bed of the gully is at a lower angle slope than the overall land slope. Discontinuous gullies erode at the upslope head, but sediment themselves at the end of the discontinuity. Hence, several discontinuous gullies may occupy the same landscape depression, their shapes progressively moving upslope. Gullies are obvious features in a landscape, and may be very large (see Plate 16 of a gully in Bolivia that is 25m wide and 15m deep at the places marked by arrows), causing the undermining of buildings, roads and trees.

How does it occur?

A gully is caused by the action of water. Runoff is channelled into grooves which deepen over time to form a distinct head with steep sides. Gullies extend and deepen in an up-valley direction by waterfall erosion and progressive collapse of their upslope parts; gully sides may collapse by water seepage or undermining by water flow within the gully.

Where do they occur?

Several conditions are conducive to gully development. They tend to form where land slopes are long and land use has resulted in loss of vegetation and exposure of the soil surface over a large area, so that the land now produces more runoff. They are particularly prevalent in deep loamy to clayey materials, in unstable clays (eg sodic soils), on pediments immediately downslope of bare rock surfaces and on very steep slopes subject to seepage of water and to landslides.

How can they be measured?

The measurement of soil loss from gullies is essentially the same as that for rills, except on a larger scale and with a different cross-sectional shape. Gullies usually have a flat floor and sloping sides, and account must be taken of these. In measuring gullies, the estimate being made is of the amount of soil displaced from the area now occupied by the gully furrows. This calculation does not include any estimate of the amount of sheet erosion occurring on the land adjacent to the gully.

In order to calculate the quantity of soil lost, it is necessary to measure the depth, width at lip and base and length of the gully (Figure 4.5). Large gullies could be measured by standard field survey equipment such as a dumpy level, although often a 30–100m tape and inclinometer are sufficient. Measurements of width and depth should be made at a number of points along the gully. If there are big variations in the width and/or depth of the gully, it is best to break the gully into similar sections and calculate the amount of soil lost for each part. These can be summed to give the total amount of soil lost from the gully.

Gullies, as more or less permanent features of the landscape, present a good opportunity to keep a time series record of their extension (or sedimentation). Repeated visits and simple measurements, plus aerial photographic and historic evidence, enable monitoring of the catchment condition over time-spans of more than 50 years. Techniques that have been used include:

- time-series aerial photography; gully progression can be directly measured;
- interviews with older members of the community and transect walks to indicate where the gully stood at significant dates in the past;

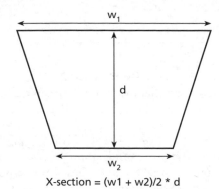

X-section = (w1 + w2)/2 * d

Figure 4.5 *Cross-sectional Area of a Trapezium-shaped Gully*

- use of permanent monitoring stakes and repeated survey measurements after major storms.

An annual rate of soil loss may not be meaningful, even if the number of years that the gully has been in existence can be established, because different rates of soil loss will occur as the gully deepens and encounters different layers of soil. The headward extension of a gully is very dependent on the condition of the catchment, sheet erosion sediment production and rates of runoff. As a gully grows, much of the soil loss may come from the sides of the gully and not sheet-wash. If catchments are conserved or planted to forest, gullies may stabilize and heal.

The volume of soil lost from a gully can be converted into an equivalent tonnes/hectare measurement, but the usefulness of this measure is limited. First, there is the same problem as with rills: what is the contributing area? Box 4.3 presents a possible choice in area for quoting soil loss – either the catchment to the gully or the field area in which the gullies are found. Second, the actual volume of the gully is only a small fraction of total sediment loss from the catchment. Third, the gully is more a symptom of a degraded catchment rather than the degradation itself.

Potential for error

1   Gullies very often visually dominate the landscape. Many conservation schemes erroneously focus on the gully, rather than the reason for the gully, which lies in the catchment. It is easy to forget that sheet erosion is likely to be ongoing and probably far greater in total sediment production.

2   Care needs to be exercised in measuring the catchment for gullies in order to make assessments of soil loss per hectare. In particular, the contributing area providing runoff decreases as the gully head extends up valley. Large gullies can be assessed from aerial photography or even maps.

Worked Example

Field Form: Gully
Site:
Date:

Measurement number	Width at lip(w_1) m	Width at base (w_2) m	Depth m
1	10.0	4.0	2.1
2	12.0	5.0	2.1
3	11.0	4.0	1.9
4	12.0	6.0	1.8
5	9.0	6.0	2.1
6	9.0	3.0	2.2
7	11.0	5.0	2.0

8	9.0	5.0	2.3
9	10.0	4.0	2.4
10	12.0	5.0	2.2
11	14.0	6.0	2.3
12	9.0	6.0	1.8
13	9.0	4.0	1.9
14	11.0	5.0	1.8
15	10.0	4.0	1.7
16	9.0	5.0	2.0
17	8.0	3.0	2.0
18	10.0	5.0	1.7
19	11.0	6.0	1.9
20	8.0	5.0	1.8
Sum of all measurements	204.0	96.0	40.0
Average (m)	WIDTH w_1 = 10.2	WIDTH w_2 = 4.8	DEPTH (d) = 2.0

Length of gully (m)　　200
Contributing (catchment) area to gully (m²)　　1,000,000

Calculations

1　Calculate the average cross-sectional area of the gully, using the formula (w1 + w2)÷2 x d.

$\boxed{^1/_2}$ (AV WIDTH w_1 +AV WIDTH w_2) $\boxed{^1/_2(10.2+4.8)}$ x DEPTH (m) $\boxed{2.0}$ = CROSS-SEC $\boxed{15}$ AREA (m²)

2　Calculate the volume of soil lost from the gully, assuming that the measurements above were taken from a gully measuring 200m in length.

CROSS-SEC AREA (m²) $\boxed{15}$ x LENGTH (m) $\boxed{200}$ = VOLUME LOST (m³) $\boxed{3000}$

3　Convert the volume lost to a per metre equivalent, assuming a catchment area of 1km², or 1,000,000m².

VOLUME LOST (m³) $\boxed{3000}$ ÷ CATCHMENT AREA (m²) $\boxed{1,000,000}$ = SOIL LOSS $\boxed{0.003}$ (m³/m²)

4　Convert the volume lost to tonnes per hectare over the whole catchment area.

SOIL LOSS (m³/m²) $\boxed{0.003}$ x CONVERSION TO t/ha $\boxed{13,000}$ = SOIL LOSS (t/ha) $\boxed{39}$

Pedestals

What are they?

A pedestal is a column of soil standing out from the surrounding eroded surface, protected by a cap of resistant material (such as a stone or root). Plate 17 shows pedestals in a field of carrot seedlings in Sri Lanka. Bunch grasses can also protect the soil immediately under it and give a pedestal-like feature – but note 'Potential for Error' 2 below. Pedestals are useful as an indicator of high sheet erosion rates of the order of 50 or more tonnes/hectare/year.

How do they occur?

Pedestals are caused by differential rain-splash erosion, which dislodges soil particles surrounding the pedestal but not under the resistant capping material (Figure 4.6). The soil particles in the pedestal itself are unaffected because they are protected by a material that harmlessly absorbs the power of raindrops. (Pedestals can be artificially simulated by using bottle tops pressed into the soil. Pedestals are created, as the bottle top protects the soil beneath from erosion, whereas the surrounding soil is exposed. They give a ready indicator to monitor, especially on surfaces where erosion rates are very large due to high intensity rainfall.)

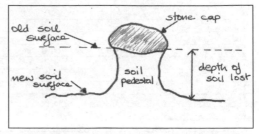

Figure 4.6 *Sketch of Soil Pedestal Capped by a Stone*

Where do they occur?
Pedestals occur on easily eroded soils, where random protection from erosion is afforded by stones or tree roots. Pedestals are often formed under trees or crops because the intercepted water falls to the ground as larger drops with greater energy to displace soil particles. The presence of gravel, pebbles or very coarse sand particles is needed, or other capping materials – but not in excessive amounts (see 'Armour Layer') as this obliterates the differential effect.

How can they be measured?
The height of pedestals can be measured using a ruler. Assuming that the cap was at the surface when erosion started, the measurement should be from the base of the stone or other capping material to the base of the pedestal, where it meets the surrounding eroded surface. The difference between the height of the pedestal and the surrounding soil surface represents the soil loss since the soil was last disturbed by tilling or other agricultural practice. Therefore, by knowing the timing of the disturbance, it is possible to estimate a rate of soil loss.

Where possible, a number of measurements should be obtained from different parts of the field. A single pedestal, or a concentration of pedestals in a particular area, are not necessarily indicative of the occurrence of sheet erosion. It is usual to take a large number of pedestal heights and express overall erosion or lowering of the ground surface as an average of these

heights. It is recommended to divide the field into a number of small areas or localities of about 1m², and take the maximum pedestal height in each locality – see 'Potential for error' 3 below.

Potential for error
1   As noted above, pedestals often form under trees or crops where intercepted raindrops fall to the ground as larger drops. If this is the only location in which pedestals are found, they would provide an unreliable estimate of the level of soil loss for a larger area.
2   Pedestals can be confused with clumps of sediment trapped by vegetation. In this instance deposition, rather than erosion, is the feature demonstrated by soil accumulations.
3   Capping stones that were originally buried in the soil may become exhumed by erosion and subsequently form a pedestal. Therefore, the height of the pedestal in such cases will underestimate erosion. It is therefore recommended that only the highest pedestals (where it may be assumed the capping material was on the surface) are taken for assessment in any small locality.
4   Material removed from around pedestals may be re-deposited elsewhere in the field. Should this occur, an estimate of the quantity of soil re-deposited must be subtracted from the calculated soil loss to arrive at the net soil loss.

Worked Example

Field Form: Pedestals

Site: Farm of Mr Nzirwehi, Bushwere, Uganda
Date: 13 February 2001

Measurement number	Maximum height of pedestal in locality (mm)
1	10
2	12
3	10
4	15
5	10
6	14
7	14
8	13
9	14
10	11
11	12
12	10
13	10
14	8
15	12
16	13
17	11
18	15
19	17
20	10
Sum of all measurements	241
Average (mm)	AV PED HEIGHT = 12.05

Calculations

1 Calculate t/ha equivalent of the net soil loss (represented by the average pedestal height).

AV PED HEIGHT (mm) $\boxed{12.05}$ x CONVERSION TO t/ha $\boxed{13}$ = SOIL LOSS (t/ha) $\boxed{157}$

Armour Layer

What is it?

An armour layer is the concentration, at the soil surface, of coarser soil particles that would ordinarily be randomly distributed throughout the topsoil. Such a concentration of coarse material usually indicates that finer soil particles have been selectively removed by erosion.

How does it occur?

Raindrops or the power of the wind detach the finer and more easily eroded soil particles. Then water or wind carries them away from the topsoil surface, leaving behind the coarser particles.

Where does it occur?

An armour layer is most likely to form on soils which have both a stony/coarse fraction as well as fine clays, silts and organic matter, following rainfall/severe winds.

How can it be measured?

Dig a small hole that shows the undisturbed armour layer. Using a ruler, measure the depth of the coarse top layer (see Plate 18). Where the depth of the armour layer is less than 1mm, it is best to scrape the stones from a small area of about three times the size and then measure this depth, and divide by three. This helps to reduce the inaccuracies in trying to measure very small depths of stones. Several measurements at different places in the field should be made in order to calculate the average depth of the armour layer.

The approximate proportion of stones/coarse particles in the topsoil below the armour layer is judged by taking a handful of topsoil from below the armour layer and separating the coarse particles from the rest of the soil. In the palm of the hand, an estimate is made of the percentage of coarse particles in the original soil. Again, this estimation should be repeated at different points in the field.

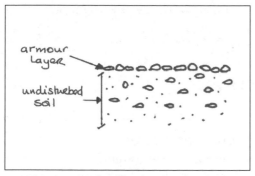

Figure 4.7 *Sketch of Armour Layer*

The depth of the armour layer is then compared with the amount of topsoil that would have contained that quantity of coarse material. The amount of finer soil particles that have been lost through erosion can then be estimated.

These calculations tell us the amount of fine particles that have been lost since the soil was last disturbed – for example, since it was tilled or weeded.

Potential for error

1   Stones on the surface may arise for other reasons, such as the exhumation of a concentration of stones in the subsurface soil.

2   The depth of the armour layer will most likely be measured to the nearest millimetre. For every millimetre the equivalent soil loss is 13t/ha (assuming an average bulk density of $1.3t/m^3$). Therefore, the accuracy of the measurements will be important in arriving at the soil loss.

3   The calculations also rely on a subjective assessment of the proportion of coarse material occurring in the topsoil. It is useful to check percentage estimates with colleagues in the field to see if there is any appreciable difference.

4   As well as the erosion process itself, repeated shallow tilling of the soil, especially in weeding operations, may concentrate more stones near the surface. Where this happens, the erosion rate will be exaggerated if the percentage concentration of stones in the original soil is based on an estimate well below the topsoil. Closer inspection of stone concentration can help to correct for this.

Worked Example

Field Form: Armour Layer

Site: Farm of Mr Karuhuru, Bushwere, Uganda
Date: 15 February 2001

Measurement number	Depth of armour layer (in mm)	Proportion of coarse material in topsoil (%)
1	0.9	20
2	1.1	25
3	1.0	15
4	1.1	22
5	0.9	20
6	1.2	20
7	0.8	22
8	0.9	19
9	1.1	20
10	1.1	20
11	1.2	18
12	1.0	20
13	0.8	18
14	0.9	22
15	0.7	22
16	1.0	20
17	1.1	18
18	1.2	20
19	1.1	20
20	0.9	19
Sum of all measurements	20.0	400
Average	AL DEPTH (mm)= 1.0	COARSE % = 20%

Calculations

1   Calculate the depth of soil required to generate the depth of the armour layer, based on the measured estimate of 20 per cent (or $^1/_5$) for coarse material in the topsoil.

AL DEPTH (mm) $\boxed{1.0}$ x COARSE % $\boxed{20\% \text{ or } ^1/_5}$ = TOTAL SOIL(mm) $\boxed{5.0}$

2 Calculate the soil lost

TOTAL SOIL (mm) $\boxed{5.0}$ – AL DEPTH (mm) $\boxed{1.0}$ = NET SOIL LOSS (mm) $\boxed{4.0}$

3   Calculate t/ha equivalent of net soil loss – using an average bulk density of $1.3t/m^3$. At this bulk density 1mm of soil loss is equivalent to 13t/ha.

NET SOIL LOSS (mm) $\boxed{4.0}$ x CONVERSION TO t/ha $\boxed{13}$ = SOIL LOSS (t/ha) $\boxed{52}$

Plant/Tree Root Exposure

What is it?

Plant or tree root exposure describes a situation where the base of the tree trunk or lateral roots are partially exposed above the present soil surface. Often a mark can be located on the trunk of a tree or stem of a plant to indicate where the original soil surface was when the tree/plant started to grow. In Plate 19 such a mark is clearly visible.

How does it occur?

Root exposure may occur where soil particles are removed by water or wind, lowering the overall soil level. Stemflow may be particularly relevant, especially where water is funnelled between exposed roots. Away from the stem, roots can act as a cap and protect against rain-splash erosion in the manner of a pedestal (see 'Pedestals').

Where does it occur?

Root exposure occurs where crops or trees are growing in areas subject to erosion.

How can it be measured?

Using a ruler, measure the distance from the soil surface to the point on the plant stem/tree trunk which would have originally been at ground level. For lateral roots away from the stem, the upper surface of the most exposed roots is usually taken as the former soil surface. Because the measurements depend on the standing plants/trees in a particular field, it may not be possible to repeat the measurements at different points throughout the field. However, if root exposure is evident at different places in the field, a number of measurements should be taken to assess the average soil loss. Differences in root exposure may reflect different erosion processes (for example, stemflow and rain-splash) occurring in the same field.

The measurement gives an estimate of the soil lost since the plant/tree was planted (but see 'Potential for error' 2 and 3, below). In the case of a crop, this may be a single growing season, whereas in the case of a tree it will depend on the age of the tree. The age of a tree is best determined by enquiry from the farmer, verified by direct observation. Independent verification of the age of many trees can be obtained by counting tree rings (known as dendrochronology). However, tree rings are not always made annually, especially in the tropics and subtropics (see Box 4.4), so it is important to know the growth patterns of any tree species used to age an erosion feature. Mature trees are sampled for their rings using a tree corer. Opportunity may also be taken if trees have been cut for fuelwood or other purposes, where it may be possible to examine a complete transverse section of the stump. The annual soil loss is calculated by dividing the measured difference between the actual soil level and that which existed when the plant/tree started to grow, by the age (in years) of the tree.

Potential for error

1   While obstacles in a field may provide indications of soil loss, these may not be representative of the soil loss in the field as a whole. The obstacle may cause channelling of erosive water flows, thus increasing the soil loss around the obstacle, or it may slow down the surface flow, allowing deposition to occur. Therefore, extrapolated soil losses, calculated solely by reference to plant/tree root exposure, may be either overstated or under-

stated. This is why it is useful to include erosion estimates from lateral roots located away from the trunk.

2   Possible errors are introduced where plants tend to heave themselves out of the ground as they grow, thereby giving a spurious impression of high soil loss. This effect is often indicated in stony soils, especially where larger platy fragments occur. Look for evidence in the alignment of stones, as tree growth may force a rearrangement of stones so that they become tilted, with the raised end nearest to the trunk. The air roots of maize plants (see Plate 20) can also be deceptive.

3   Related to the above error is the expansion of root diameter as the tree grows. Roots running parallel to the surface may rise to/above soil level. This gives the appearance that there has been more erosion than has actually occurred.

BOX 4.4 CHOOSING SUITABLE TREES

Both the techniques of using tree mounds and those of examining tree root exposure demand certain characteristics of the Indicator tree species. First, in its growth cycle the tree must not do anything unusual, such as heave itself out of the ground as the roots enlarge. Another unusual phenomenon noted with gum (*Eucalyptus* spp.) trees in parts of Africa is piping and underground erosion, to the extent that the tree itself may subside into the hole so created. Second, the tree should be capable of being dated. In farmers' fields, trees may well have been planted and dates are easy to determine from the land user. But elsewhere the assessor has to rely on dendrochronology or counting tree rings. Only some 10–30 per cent of tropical trees have clearly defined growth rings.

A full listing cannot be given here (see Humphreys and Macris reference in Appendix IV) but trees that have been used successfully in various parts of the world to assess land degradation include:

- *Acacia drepanalobium* – the ant-gall acacia, a slow-growing thorny shrub
- *Acacia albida* – throughout Africa planted in fields because of its beneficial shade and leaf fall
- *Tectona grandis* – teak, very tall with large leaves, planted on a 40+ year cycle. Erosion is often high underneath the canopy and roots become exposed easily
- *Tamarindus indica* – the tamarind, planted as a long-lasting fruit tree throughout the tropics. Its roots easily become exposed
- *Parinari curatellifolia* – a fruit tree (the fruit used for wine) that grows on problematic sodic soils in Africa. Farmers usually retain these magnificent trees In their fields and their age will be known by oral tradition

Trees with buttresses, common in rainforest environments, are not suitable for the tree root exposure technique. The buttress does not imply that the original soil level has lowered.

Worked Example

Field form: Plant/Tree Root Exposure

Site: Farm of Mr Kakuba, Bushwere, Uganda
Date: 15 February 2001

Measurement number	Measured difference in soil level (A) mm	Converted to tonnes/hectare A x 13 t/ha	Age of plant/tree yrs	Annual change in level t/ha/yr
1	7	91	5	18.2
2	6	78	5	15.6
3	7	91	5	18.2
4	8	104	5	20.8
5	8	104	5	20.8
6	6	78	4	19.5
7	3	39	2	19.5
8	2	26	2	13.0
9				
10				
11				
12				
13				
14				
15				
16				
17				
18				
19				
20				
Sum of all measurements	–	–	–	145.6
Average t/ha/yr	–	–	–	ANNUAL SL = 18

Note: In this case it has been assumed that coffee plants of different ages in the same plot have demonstrated root exposure. There is no evidence of increasing or decreasing rates of soil loss over time. So, an average soil loss rate of 18 t/ha/yr is justified.

Exposure of Below Ground Portions of Fence Posts and Other Structures

What is it?

Sheet erosion, resulting in a reduction in the general ground level, can be identified where the below ground portions or foundations of man-made structures, such as fence posts, other poles, old tracks and roads, bridges and buildings, are exposed. (Soil accumulations are also possible around these kinds of structures. Where this occurs, the technique described to quantify soil loss evidenced by build-up against barriers can be used to estimate erosion rates.)

How does it occur?

The action of wind or water detaches soil particles from the soil surface and transports them to be deposited elsewhere. Over time, assuming that removal exceeds deposition, this will lead to a reduction in the level of the soil surface. Structures with known foundations/depths below surface can be used to measure the general lowering of the soil surface.

Where does it occur?
Measurements of soil loss using man-made structures can only be made where it is clear that factors other than erosion (for example, construction) have not been the cause of soil loss. Fence posts and other poles are particularly useful since the insertion into the ground involves minimal disturbance. Track, road and bridge construction often involves a greater degree of initial disturbance, but this can be compensated for by allowing an initial settling-down period. In particular, bridges and tracks/roads often become useful markers if abandoned or no longer maintained. Buildings are a more complicated item to deal with and a great deal of care is required.

How can they be measured?
The measurement strategy clearly depends on the object used for establishing the original ground level. For fence posts and poles this can be established by determining the height of the exposed part of the post/pole and/or the length buried into the ground. Often standard post/pole lengths are used in the area (see Box 4.5). If not, it is neces-

sary to determine a typical value by measuring the above ground length of posts in those sites that appear to have been least affected by soil erosion. The distance between the new ground surface and the point on the post that would originally have been at ground level can be measured using a ruler. In some instances erosion may remove soil equivalent to the depth of the below ground portion of the post, in which case, providing it is certain that the post was not broken and that no part remains below ground, a minimum rate of erosion can be estimated. In other cases, the post may be entirely free of the soil but held in position by taut wire and hence the full extent of erosion can be measured. (Excavation is required where the post is expected to be completely buried as a result of deposition of soil particles.)

The same procedure applies to buildings and other structures, although because they are point locations they cannot provide the spatial record that fence lines can. In the case of paved paths and roads it is common for the pavement to seemingly disappear over time. This may occur by burial during sedimentation or, in areas of active erosion, especially on steep slopes, the pavement may be undermined and the path breaks up. (Such break-up can also be due to the removal of

---

BOX 4.5 EXAMPLE OF USE OF FENCE POSTS TO DETERMINE SOIL LOSS: DEGRADED RANGELANDS, AUSTRALIA

In an Australian study, straight fence lines crossed hilly terrain and floodplains to enclose grazing areas. The posts were of uniform length and each contained three uniformly positioned drill holes for fence wire. In this case the height of the lower hole was positioned at a set height above the ground surface. This configuration provided a good basis for establishing the position of the original ground surface. Erosion was indicated by an increased distance between the present day ground level and the height of the lower hole, which was measured by a ruler to the nearest 5mm. The study was able to use a number of fence lines to show a distinct spatial pattern to erosion and sedimentation and delimit areas where degradation of this type was not evident.

Source: Geoff Humphreys, Macquarie University, personal correspondence

soil from beneath the pavement by various animals such as ants, earthworms and termites.) An absence of path destruction can indicate reasonable stability and low erosion rates (Plate 21).

Potential for error

1   The use of fence posts, poles and similar structures is only possible where the age of the structure is known and where it is possible to determine the level of the original ground surface. Not being able to satisfy either parameter means that the technique is unsuitable.

2   Fences and other structures may actively promote erosion or sedimentation. Hence, fences with a railing or other barrier at or close to the ground level are better treated as an example of 'Build-up against Barriers', especially when aligned across the slope. The best fences are those without a barrier near ground level and aligned perpendicular to slope. Local scouring around a post can usually be detected since it creates a depression, often semi-circular, below the existing surface (see 'Waterfall Soil Loss'). This extra depth can be excluded from any calculation.

Worked Example

Field Form: Fence Post Exposure

Site: Australia
Date:

Measurement number	Depth of erosion (A) mm	Converted to tonnes/hectare A x 13 t/ha	Time elapsed since structure installed yrs	Annual soil loss t/ha/yr
1	20	260	45	5.8
2	55	715	45	15.9
3	40	520	45	11.6
4	105	1365	45	30.3
5	60	780	45	17.3
6	55	715	45	15.9
7	80	1040	45	23.3
8	35	455	45	10.1
9				
10				
11				
12				
13				
14				
15				
16				
17				
18				
19				
20				
Sum of all measurements	–	–	–	130.2
Average t/ha/yr	–	–	–	ANNUAL SL = 16

Note: In this example there is a substantial difference in the assessed soil loss between the posts measured. Greater accuracy would be achieved if more field posts were measured.

'Waterfall' Soil Loss

What is it?

Waterfall soil loss describes a depression or hole found on the immediate downslope side of a plant or other obstruction (see Plate 22). When first identified as a distinctive soil loss indicator in the validation workshop in Uganda, this feature was called a 'scoop' – presumably after the indentation in a tub of ice cream after a scoop has been taken. It usually has a steep cut on the upslope side forming the 'waterfall' and a low angle base to join the general slope (Figure 4.8).

How does it occur?

The plant or other point obstruction causes water to concentrate around the stem or pole during times of heavy rain and overland flow on steep slopes. This concentrated flow may be augmented by stemflow – that is, water running down the stem of the obstruction – especially if the obstructing stem has leaves which concentrate water (eg banana plants and agaves). When the flow meets again on the downslope side of the plant or obstruction, a waterfall- or eddy-effect is created that has such concentrated energy as to erode out a substantial 'scoop' of soil.

Where does it occur?

Waterfall soil losses/scoops are most likely to be found on steep slopes, particularly where soils are easily eroded and crop cover is poor. The existence of sizeable stems of perennial plants or fence poles is necessary to concentrate the water flow lines to the extent that such erosion may occur.

How can it be measured?

The volume of soil lost from the slope is represented by the soil 'missing' from the scooped-out holes. The measurement compares the surface level of the surrounding land with the level of the 'scooped'-out holes. The formula for calculating the volume of a cone is used to estimate the amount of soil lost from the scoop. In order to use this formula, the depth (at the deepest point of the scoop) and the diameter of the scoop are measured. If there are only a small number of scooped-out holes in the field then, by measuring each hole, the soil loss for the field by this process can be established. However, where there are a large number of scooped out holes in the field (for example, below every plant stem), a representative sample of measurements from different depressions throughout the field should be obtained. An average volume of soil loss can be estimated from these measurements. The number of depressions in the field should also be counted (or estimated). The average volume of soil lost from these scooped-out depressions is then multiplied by the total number of depressions in the field to get an estimate of the amount of soil lost from the field. This can then be converted into a tonnes/hectare equivalent. The rate of soil loss can be estimated by determining when

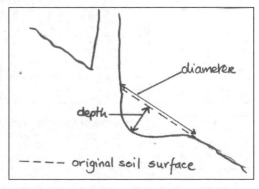

Figure 4.8 *Sketch of Depression Caused by Waterfall Effect*

the soil surface was last disturbed to the extent that the old 'scoops' had been obliterated – eg planting or weeding.

Potential for error

1   Soil loss rates by this process may be under-estimated, since these calculations do not include sheet erosion, which is likely to reduce the level of the soil surface against which the depth of the scooped-out depression is compared.

2   Harvesting of crops intercropped with standing crops may leave depressions in the soil surface. These may be mistaken for gouged-out depressions, or may contribute to the depth of these depressions.

3   Similarly, weeding around perennial stems may cause features similar to 'scoops'. On steep slopes with hand implements, farmers typically cultivate the soil downhill away from the stem. Careful inspection and participant observation with the farmer can help to avoid this error.

Worked Example

Field Form: Waterfall Effect

Site: Farm of Mr Nzirwehi, Bushwere, Uganda
Date: 13 February 2001

Measurement number	Scoop diameter m	Scoop radius (Diameter ÷ 2) r m	Scoop depth d m	Scoop volume ($^{1}/_{3} \pi \times r^2 \times d$) m³
1	0.12	0.060	0.60	0.0023
2	0.14	0.070	0.54	0.0028
3	0.14	0.070	0.50	0.0026
4	0.15	0.075	0.63	0.0037
5	0.16	0.080	0.72	0.0048
6	0.17	0.085	0.70	0.0053
7	0.15	0.075	0.65	0.0038
8	0.17	0.085	0.80	0.0061
9	0.15	0.075	0.71	0.0042
10	0.16	0.080	0.64	0.0043
11	0.16	0.080	0.64	0.0043
12	0.15	0.075	0.58	0.0034
13	0.17	0.085	0.75	0.0057
14	0.14	0.070	0.73	0.0037
15	0.14	0.070	0.79	0.0041
16	0.15	0.075	0.65	0.0038
17	0.13	0.065	0.72	0.0032
18	0.16	0.080	0.78	0.0052
19	0.17	0.085	0.66	0.0050
20	0.18	0.090	0.60	0.0051

Total volume of soil lost from these measurements (m³)	0.0834
Average volume of soil lost from each scoop (m³)	0.0042

Field area(m²)	300
Number of Scoops	212

Calculations

1 Calculate the volume of soil loss for the whole field, assuming that there are 212 scoops in the field.

AVERAGE VOLUME (m³) $\boxed{0.0042}$ x NO. OF SCOOPS $\boxed{212}$ = VOLUME OF SOIL LOST FROM FIELD (m³) $\boxed{0.8904}$

2 Calculate the soil loss per square metre, assuming the field measures 300m².

VOLUME OF SOIL LOST FROM FIELD (m³) $\boxed{0.8904}$ ÷ AREA OF FIELD (m²) $\boxed{300}$ = VOLUME OF SOIL LOST PER M² (m³/m²) $\boxed{0.0030}$

3 Calculate the tonnes per hectare equivalent of this volume of soil loss.

TOTAL VOLUME OF SOIL LOST (m³/m²) $\boxed{0.0030}$ x CONVERSION TO t/ha $\boxed{13,000}$ = SOIL LOST (t/ha) $\boxed{39}$

Rock Exposure

What Is It?

Rock exposure describes the situation where underlying rock has, because of erosion, been exposed at the ground surface.

How does it occur?

Rock exposure occurs where the soil particles that had previously overlain the rock have been removed by the action of wind or water. The bare rock surface is exhumed: ie its relative position with respect to the soil surface has changed.

Where does it occur?

Rock exposure occurs where there are shallow soils covering rocky but massive parent material. Well-weathered parent rock is not suitable because it is also erodible and fails to give a clear marker against which to measure soil removal.

How can it be measured?

It is necessary to assess where the bare rock was when accelerated erosion started to happen. This can be difficult, but two practical field situations can be considered. First, the rock may have been only partially buried. In this case, the old surface soil usually stains the rock and the previously buried part of the rock can be clearly seen. In other situations the older exposed rock is covered in lichens, whereas the recently exposed rock is not. Even when the lichens have been removed, evidence of their presence remains in the form of etching patterns on the rock surface. Measurement, then, is straightforward: take the depth of soil removed by measuring vertically from the current soil surface up to the boundary of the stained part of the rock.

Second, the rock may have been completely buried. This is likely when there are no clear staining marks to differentiate from unstained areas. In this case, the conservative assumption is that the rock will have been just below the old surface of the soil. Under this assumption the removal of soil by erosion amounts to a depth equivalent to the whole height of the rock exposure. It will be clear that this will give a minimum estimate of long-term soil loss. At least there will be no danger of spurious exaggeration! It is recommended that a large number of such measures be undertaken in order to reduce individual sampling errors. If periodic measurements are planned, markers could be left to show the current level of soil. Masonry nails provide a useful marker in harder rocks. Future measurements could then be made by reference to these markers. However, this is unlikely to yield measurable results in less than several years of erosion – in other words, after enough erosion to cause a lowering of several millimetres.

Potential for error

1 The main source of error is identifying where the rock sat in relation to the ground surface before significant erosion started. The best situations are on dark-coloured rocks which weather to rich-brown clays or lighter rocks that weather to reddish clays – in these cases, the staining is easily visible and may last for many decades. The best check is to look at a number of rock exposures and gain an overall consistent set of measures of soil removal – then greater confidence can be gained in the accuracy of the estimates.

2 The baseline of the level of the current soil surface can be problematic. The rock exposure itself alters the local hydrology and may partially protect the soil on the upslope side. On the downslope side, eddies in the water flow may cause greater scouring, thereby lowering the current soil surface more than if the rock had not been there (see 'Waterfall Soil Loss'). Only careful site inspection can confirm these sources of error. But once detected, they can be compensated for.

3 The time over which rock exposure has occurred can be difficult to estimate. Normally, it is sufficient to enquire when the land was first opened up for agriculture, and use this date to calculate the length of time over which erosion has occurred. However, the land degradation hazard may not have been evenly distributed over that time. So, results should be presented as long-term mean estimates.

4 Falling trees (or windthrow) often result in the occurrence of stones and bedrock fragments on or near the surface. Thus, especially where there is evidence of forest clearance or recently fallen trees, stones at or near the surface may not be indicative of erosion processes.

Worked Example

See 'Solution Notches'.

Solution Notches

What are they?

This is a particular but very useful case of the previous indicator, rock exposure. Solution notches are indentations found on rocks that indicate the historic soil level.

How do they occur?

Solution notches arise because of chemical reactions between the soil, air and the rock. Topsoils have greater chemical reaction with rock because of humic acids released by organic matter as it decomposes and the greater abundance of soil flora and fauna. Therefore, in the zone of the former topsoil, especially at the interface between topsoil and atmosphere, there is greater weathering of adjacent bare rock. This weathering leaves a horizontal solution mark or notch, which is often smoother than the exposed rock. Where the soil is subsequently removed (by erosion or other means), these notches become visible as permanent markers of where the soil was (see Plate 23).

Where do they occur?

Solution notches are most likely to occur on limestone and calcareous rocks. These are the rocks most susceptible to solution by acid organic chemicals, but the same effect can occur on other rocks.

How can they be measured?

Measurement is broadly similar to rock exposures. The distance from the solution notch to the current soil level gives an indication of how much soil has been eroded. This distance can be measured using a ruler, and then converted into a soil loss per hectare equivalent. A problem in using solution notches to determine a rate of soil loss is finding another indicator for calibration, so that the period over which the

soil loss has occurred can be estimated. Other indicators might include tree root exposure on nearby trees or marks of soil level on houses built a known number of years previously. Alternatively, it may be possible to date the 'exposed' surface by reference to solution pitting, but since this occurs at a rate of 2–5mm/1000 years the surface must have been exposed for at least a few hundred years.

Potential for error

1 Marks on rocks may not be solution notches. They may indicate other forms of damage, for example scraping by machinery. However, this type of damage initially produces rough, broken surfaces with sharp edges in contrast to the smooth form of natural notches.

2 The amount of soil loss around the base of a rock may be less, or greater, than that which occurs nearby. Deposition of soil particles may occur against the rock (as in build-up against barriers) or channels may form around the base of the rock, increasing the amount of soil loss.

3 It can be difficult to determine an appropriate time-span over which the erosion occurred.

Worked Example

This example applies to both rock exposures and solution notches. A large number of individual measurements should be carried out in the locality, and results compared to gain a view of the consistency of the evidence. Other techniques such as root exposure and tree mounds should be undertaken to corroborate the results from exposed rocks.

Field Form: Solution Notches/Rock Exposure

Site:
Date:

Measurement number	Distance of solution notch/exposure from ground surface (mm)
1	224
2	222

3	216
4	210
5	227
6	220
7	221
8	
9	
10	
11	
12	
13	
14	
15	
16	
17	
18	
19	
20	
Sum of all measurements	1540
Average (mm)	AV HEIGHT = 220

Calculations

1 Calculate annual change in the soil level, assuming that the land was deforested and agriculture commenced 20 years previously.

AV SOIL LOSS (mm) $\boxed{220}$ ÷ PERIOD OF SOIL LOSS (yr) $\boxed{20}$ = ANNUAL SOIL $\boxed{11}$ LOSS (mm/yr)

2 Calculate the t/ha equivalent of the annual soil loss.

ANNUAL SOIL LOSS (mm/yr) $\boxed{11}$ x CONVERSION TO t/ha $\boxed{13}$ = SOIL LOSS (t/ha/yr) $\boxed{143}$

Tree Mound

What is it?

A tree mound describes the situation where the soil under a tree canopy is at a higher level than the soil in the surrounding area (illustrated in Plate 24). A tree mound has approximately the same diameter and shape as the overhanging tree canopy.

How does it occur?

The presence of tree mounds indicates that there has been more erosion away from the tree than near it, since the surface of the mound represents an earlier soil level. Hence, it can be concluded that the erosive impact of the raindrops is absorbed by the tree canopy. This reduces the eroding power of raindrops reaching the ground surface, and therefore the amount of soil dislodged. In contrast, soil unprotected by a tree canopy is subject to the full force of the raindrops, so that soil particles are dislodged and are transported downslope in runoff. Thereby, a local difference in soil surface level is constructed, enabling field assessment of historical erosion rates in the general area as compared with the baseline level of minor erosion under the tree.

Where does it occur?

Tree mounds occur where a tree provides good, continuous protection to the ground surface (Figure 4.9). The best sites for assessment are on extensive low-angle plains of semi-arid zones, where occasional trees dot the landscape. The original area where this technique was developed was in East Africa with the ant-gall acacia, *Acacia drepanalobium* (see Box 4.4).

How can it be measured?

The level of the soil surface under the tree and in the open is compared. The difference in height between the soil surface under the tree and in the surrounding area gives an approximation of the soil loss that has occurred during the life of the tree.

The life of the tree can be assessed by asking local people, or by dendrochronology. The non-destructive approach is to be preferred. However, in the original area where this technique was developed, a selected number of ant-gall acacias were felled, their age assessed from rings in the trunk, and their circumference measured. Regression graphs were then constructed of tree age versus circumference. These calibration graphs then enabled age assessment to be undertaken non-destructively for all further sites in the local area.

A large number of measurements should be taken. First, the height difference between maximum elevation of the mound and general level of soil surface should be measured. A spirit level is useful here to extrapolate from the top of the mound horizontally away from the tree to the general soil surface. The difference in soil level close to the edge of the tree canopy can be easily measured using a ruler.

Second, the historical erosion rate should be calculated and the time period noted over which this historical rate applies. Third, the measurements are repeated for trees of different sizes and ages. It is recommended that erosion rates be grouped according to bands of progressively older time periods to see if there is any difference in calculated average erosion rate. Typically, in the East African case, the longer the time period, the lower was the average erosion. This indicated that erosion rates have been increasing in recent historical times as grazing pressures have increased.

Potential for error

1   Mounds around the base of trees, shrubs and other plants may have been caused by factors other than erosion. Such factors include the binding effect of the tree roots, the greater incorporation of organic matter into the soil beneath the tree, the displacement of soil during tree growth and the construction of nests by ants and termites. Also, some have claimed that livestock sit under trees during the day for shade, thereby providing greater inputs of manure. In addition, the trunks of trees may act as a barrier to the transport of sediment, resulting in deposition. Organic matter may build up under trees, especially where leaf litter accumulates or livestock shelter.

2   Some trees may lift the soil around them as they grow, thus creating natural mounds and an appearance of higher levels of soil loss than may actually exist.[1]

3   Because the tree canopy size changes as the tree grows, the tree mound will not be at a constant height above the level

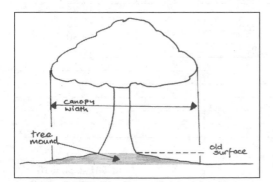

Figure 4.9 *Sketch of Tree Mound*

of the surrounding soil. Thus, it is important to take measurements at different points from the edge of the mound towards the tree trunk.

4   Wind-borne sediment can be slowed or trapped by trees and shrubs, falling to the ground surface underneath the leaf canopy. Such material increases the difference between the surface beneath the tree and beyond, but it bears no relation to the original soil level and may have been transported from far off.

5   Counting tree rings as an estimate of age of tree is problematic in many tropical species. Rings may occur seasonally, where there are two rainy seasons per year as in East Africa. They may merely indicate longer cycle climatic conditions such as runs of wet and dry years. Careful checking is needed, and local advice and information from farmers is invaluable. On cropland, trees will usually have been planted and the land user will be able to give direct information (see Box 4.4).

6   Sometimes farmers deposit organic refuse around the base of trees when weeding garden plots. This is particularly common in humid areas, such as the highlands of Papua New Guinea. Alternatively, farmers may remove the tree mound, especially if it contains organic matter, to distribute to their fields.

Worked Example

Field Form: Tree Mound
Site:
Date:

Measurement number	Measured difference in soil level (A) mm	Converted to tonnes/hectare A x 13 t/ha	Age of plant/tree yrs	Annual soil loss due to change in surface level t/ha/yr
1	35.0	455.0	25	18.20
2	28.0	364.0	20	18.20
3	22.5	292.5	18	16.25
4	18.0	234.0	10	23.40
5	21.0	273.0	15	18.20
6	24.0	312.0	15	20.80
7	21.0	273.0	15	18.20
8	22.5	292.5	18	16.25
9	27.5	357.5	22	16.25
10	27.5	357.5	22	16.25
11	29.0	377.0	20	18.85
12	27.0	351.0	20	17.55
13	22.5	292.5	15	19.50
14	22.5	292.5	18	16.25
15				
16				
17				
18				
19				
20				
Sum of all measurements	–	–	–	254.15
Average (t/ha/yr)	–	–	–	ANNUAL SL = 18

Note: In this case it has been assumed that trees of different ages in the same plot have demonstrated tree mounds

Build-up against Barriers

What is it?

Where the transport of eroded material is halted by an obstruction, the particles suspended in the runoff may be deposited against the obstruction as the water slows. This results in a build-up of sediment against the barrier (as illustrated in Plate 25). This indicator measures movement of soil across the field rather than loss from the field.

How does it occur?

Fine soil particles are transported in water. If the runoff meets a barrier, its speed is reduced and soil particles settle out of suspension, thereby creating a small sedimentary layer. On steeper slopes, and especially when the soil is dry, clods of soil may roll downslope with the slightest disturbance. Over time, such deposited matter will alter the slope surface. This build-up is often accelerated by plough erosion – see 'Potential for error' 3.

Where does it occur?

Build-up against barriers will occur where an obstruction exists to bar the transport of fine soil particles. Typical obstructions are field boundaries, logs on the surface, stone bunds and fence lines.

How can it be measured?

The volume of soil trapped behind the barrier is calculated by measuring the depth of the soil deposited and the area over which it is deposited (see Figure 4.10). Where the build-up is against a continuous barrier such as a fence or hedge, the measurement will give an approximation of soil loss from the field.

A visual examination of the area close to the barrier will indicate how far the deposition extends into the field. This distance (length) should be measured at a number of points. The depth of the soil accumulated against the barrier can be determined by examining the soil level against the

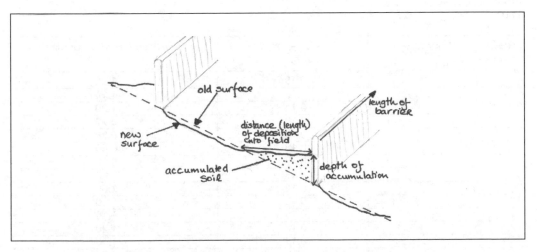

Figure 4.10 *Sketch of Build-up of Eroded Material Against a Barrier*

barrier on the other side from the accumulation. (There is a danger that because of soil erosion on the lower field, the soil level next to the barrier will have been lowered.) As illustrated in Plate 25, the depth of the accumulation of soil is not constant. In order to calculate the amount of soil accumulated, a linear slope is assumed.

The amount of soil accumulated behind a barrier represents a build-up over time. The annual rate of soil loss from a hillside is arrived at by dividing the quantity of accumulated soil by the number of years that a barrier has been in existence. Older barriers can be treated as archaeological sites and careful excavation can lead to the recovery of datable materials such as charcoal and artefacts.

Potential for error

1   The calculations do not differentiate between sediment that results from in-field erosion and sediment that results from erosion further upslope and outside the immediate field, which may lead to an overestimation of the soil loss per field.

2   Not all materials transported in runoff will be deposited at a barrier. The speed, volume and direction of runoff all influence the level of deposition. Therefore, the estimated soil loss may be understated by the amount of soil carried beyond the barrier.

3   Tillage techniques may increase the soil depth behind barriers, particularly where conservation techniques such as terracing have been introduced to lessen the effect of slope. This tillage erosion is also called 'plough' erosion, because farmers often scrape soil downhill when they cultivate.

4   If the slope were convex before the barrier was constructed, the estimate of soil loss will be understated, as it assumes a linear slope.

Worked Example

Field Form: Build-up against Barrier

Site:

Date:

Measurement number	Measured depth m	Measured length of deposition m
1	0.18	1.00
2	0.12	1.10
3	0.14	1.20
4	0.19	0.70
5	0.18	0.80
6	0.18	0.60
7	0.17	0.90
8	0.13	0.90
9	0.14	1.00
10	0.15	1.20
11	0.15	1.10
12	0.12	1.20
13	0.19	1.00
14	0.19	0.80
15	0.14	0.70
16	0.16	0.90
17	0.15	0.70
18	0.17	1.00

19	0.17	1.10
20	0.18	1.00
Total	3.2	18.9
Average(m)	0.16	0.95

	Length of barrier (m)	7.00
	Contributing (catchment) area to barrier (m²)	70.0

Calculations

1 Calculate the average cross-sectional area of the accumulation, using the formula for the area of a triangle (ie $\frac{1}{2}$ base x height), where the base equates to the measured length of the deposition from the barrier into the field and the height equates to the depth of the deposition.

$\boxed{\frac{1}{2}}$ x DEPTH (m) $\boxed{0.16}$ x LENGTH (m) $\boxed{0.945}$ = CROSS-SEC AREA (m²) $\boxed{0.07560}$

2 Calculate the volume of soil accumulated behind the barrier, assuming that the barrier measures 7m in length.

CROSS-SEC AREA (m²) $\boxed{0.07560}$ x BARRIER (m) $\boxed{7}$ = VOLUME ACCUMULATED (m³) $\boxed{0.5292}$

3 Convert the total volume accumulated to a volume per square metre of contributing area, of 70m².

VOLUME ACCUMULATED (m³) $\boxed{0.5292}$ ÷ CONTRIBUTING AREA (m²) $\boxed{70}$ = SOIL LOSS $\boxed{0.00756}$ (m³/m²)

4 Convert the volume per square metre to tonnes per hectare.

SOIL LOSS (m³/m²) $\boxed{0.00756}$ x CONVERSION TO t/ha $\boxed{13,000}$ = SOIL LOSS (t/ha) $\boxed{98.3}$

5 Convert the total soil loss as represented by the soil accumulated behind the barrier into an annual equivalent, assuming that the barrier was constructed 3 years before the measurements were recorded.

SOIL LOSS (t/ha) $\boxed{98.3}$ ÷ TIME (yr) $\boxed{3}$ = ANNUAL SOIL LOSS (t/ha/yr) $\boxed{33}$

Build-up against Tree Trunk/Plant Stem

What is it?

This indicator occurs as an accumulation of soil on the upslope side of a tree trunk or plant stem (see Plate 26). It is closely related to the previous technique, 'Build-up against Barriers', but the shape of the accumulation is different. In the Ugandan validation workshop, it was seen to be significantly different from the build-up against a linear barrier.

How does it occur?

The tree trunk or plant stem acts as a barrier to runoff from the slope. The trunk or stem slows down runoff. Heavier particles carried in the runoff are deposited against the barrier. The deposition is usually aided by the trunk or stem trapping organic matter such as leaves, which in turn then filter out sediment to accumulate behind the barrier.

Where does it occur?

Build-up against tree trunks or plant stems will occur on slopes where the trees/plants act as a barrier to runoff. This is usually on slopes of more than 20 per cent – on lower angle slopes, the tree mound technique is normally more appropriate.

How can it be measured?

The volume of soil accumulated is calculated by measuring the depth of the accumulation (at the deepest point) and the distance that it extends from the tree trunk or plant stem. For ease of calculation it is assumed that the shape of the accumulation is a regular, half-cone shape (see Figure 4.11). If the distance from the trunk/stem to the edge of the deposited materials is not reasonably constant, several measurements should be taken and

the average of these measurements used for the purpose of these calculations.

The accumulated soil on the upslope side of the tree or plant represents a build-up over time. The length of time that it has taken to accumulate the visible deposition will depend on how long the tree or plant has been growing in the field, and on land management practices. This information should be obtained from the farmer.

The measurement of soil accumulated behind tree trunks or plant stems gives an estimate of the eroded material 'saved'. However, eroded material will have been lost in runoff occurring between the trees or plants. The annual rate of accumulation behind tree trunks and plant stems can be assumed to equate, on a conservative basis, to the rate of soil loss for the remainder of the field. This indicator is good evidence for the amount of sediment moving across a slope. In pine and eucalyptus plantations it has been found that overland flow and sediment loss has been substantial – so this is a good indicator of this, as well as giving a very cautious (conservative) estimate of amounts of sediment moving across the slope.

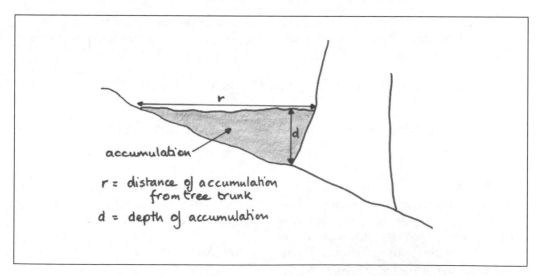

Figure 4.11 *Sketch of Accumulation Behind Tree Trunk*

Potential for error

1 Where deposition occurs, it is the coarser particles that are deposited first. Thus the accumulations may reflect only part of the soil loss from the field, since finer particles may remain suspended in the runoff and be transported off-site.

2 The 'saved' material may include deposition of materials eroded from areas further up the slope, resulting in an over-estimation of the soil loss from the field.

3 Land management practices may affect the soil level around trees and plants.

Farmers may make mounds around the trunk or stem for a number of reasons (eg to concentrate organic matter close to the tree/plant; to increase the moisture available to the plant). Alternatively, farmers may redistribute soil accumulated above trees and plants so that any visible accumulation may relate to a much shorter period than the life of the tree/plant.

4 Termite mounds have sometimes been mistaken for accumulations around a tree trunk – careful inspection will usually determine this.

Worked Example

Field Form: Build-up against Tree Trunks/Plant Stems
Site: Farm of Mr Kakuba, Bushwere, Uganda
Date: 13 February 2001

Measurement number	Depth d	Distance from trunk/stem r	Volume of soil saved $^1/_2(^1/_3 \pi \times r^2 \times d)$	Contributing area
	m	m	m³	m²
1	0.19	0.13	0.0017	0.15
2	0.17	0.12	0.0013	0.24
3	0.16	0.11	0.0010	0.20
4	0.16	0.12	0.0012	0.22
5	0.18	0.13	0.0016	0.30
6	0.19	0.14	0.0020	0.35
7				
8				
9				
10				
11				
12				
13				
14				
15				
16				
17				
18				
19				
20				

Total Volume Saved (m³)	0.0087
Total Contributing Area (m²)	1.46
Age of Trees	4 years

Note: For the purposes of this example it has been assumed that there are only six trees in the farmer's field, all planted four years previously, which demonstrate evidence of accumulations on the upslope sides.

Calculations

1   Calculate the annual rate of soil accumulation, based on trees aged four years.

TOTAL VOLUME OF $\boxed{0.0087}$ ÷ AGE OF TREES $\boxed{4}$ = ANNUAL VOLUME OF $\boxed{0.0022}$
SOIL SAVED (m³)   (YEARS)   SOIL ACCUMULATED (m³/yr)

2   Convert the total volume of soil accumulated to a volume per square metre.

ANNUAL VOLUME $\boxed{0.0022}$ ÷ CONTRIBUTING $\boxed{1.46}$ = TOTAL VOLUME $\boxed{0.001507}$
OF SOIL   AREA (m²)   OF SOIL
ACCUMULATED (m³/yr)   ACCUMULATED (m³/m²)

3   Calculate the tonnes per hectare equivalent of this volume of soil accumulated (and, thus, the tonnes per hectare equivalent of soil lost between the tree/plant barriers).

TOTAL VOLUME $\boxed{0.001507}$ x CONVERSION TO t/ha $\boxed{13,000}$ = SOIL LOST $\boxed{20}$
OF SOIL ACCUMULATED/   (t/ha/yr)
LOST (m³/m²)

Sediment in Drains

What is it?

On agricultural land, runoff from a hillside is often channelled off the slope via drains running across the slope that are designed to protect the land from excess runoff. Sediment being carried in the runoff may be deposited as the water passes along the drains. Plate 27 shows fine material deposited in a furrow in a field in Venezuela.

How does it occur?

As runoff slows down on entering the across-slope drain, the eroded materials being carried are deposited within the drain. The process is exactly the same as sedimentation in a riverbed, where the velocity of the flow ceases to be sufficient to carry particles in suspension. The deposited sediment indicates the amount and type of material that has been eroded from the land above the drain.

Where does it occur?

Sediment deposition occurs in most places where erosion occurs, as particles of soil dislodged are inevitably re-deposited else-where downslope – in this case in drains which act as sediment traps.

How can it be measured?

The difference between the surface level of the drain before and after deposition represents the quantity of eroded material deposited from the drain's catchment area. The sediment in the drain can be measured by calculating the depth of the sediment, the width and length of the drain. By multiplying these three figures together, the volume of soil deposited in the drain can be estimated. A number of measurements at different points along the drain should be taken to obtain an average depth of sediment deposited and an average width of the drain.

Potential for error

1   Run-on to an area carries sediment. If deposited in the drain, this sediment is measured as if it had come from the drain's catchment area rather than from further upslope, thus resulting in a possible overstatement of the amount of soil loss.

2 Eroded material that is very fine (such as organic matter, clays and silts) may not be deposited in the drain but be deposited further downstream. This eroded material is completely missed by these calculations. This means that the amount of erosion from a plot may be understated, particularly if the greatest soil loss occurs after a small number of large rainfall events/storms, rather than continuously throughout a season.

3 The type of eroded material is also unrepresentative of total soil loss – but see 'Enrichment Ratio'.

4 Eroded material in a drain can itself be picked up by runoff in the drain and carried further downstream. Thus measurements taken after a storm event might suggest less soil loss than measurements taken in the same place before the storm.

5 In-field erosion and deposition is disregarded. Therefore, provided the eroded material does not leave the plot and is not deposited in the drain, it will not be included in this measurement of erosion.

Worked Example

Because of the potentials for error noted above, sediment in drains will tend to give a very conservative estimate of soil loss from fields. Actual values of soil loss can be estimated by multiplying by an assumed enrichment ratio – but this is not shown in the example below.

Field Form: Sediment in Drain
Site:
Date:

Measurement number	Depth of sediment mm	Width of drain mm
1	26	300
2	29	285
3	26	303
4	27	298
5	30	284
6	27	274
7	32	301
8	30	275
9	28	308
10	28	297
11	28	302
12	30	286
13	30	280
14	25	270
15	29	272
16	27	290
17	27	286
18	32	297
19	36	299
20	33	293
Sum of all measurements	580	5800
Average (mm)	DEPTH = 29	WIDTH = 290

	Length of drain: (m)	10
	Contributing (catchment) area to drain: (m²)	50

Calculations

1   Convert the average depth and width of the sediment in the drain from millimetres to metres (by multiplying by 0.001). Thus, an average depth of 29mm is equal to 0.029m and an average horizontal width of 290mm is equivalent to 0.29m.

2   Calculate the average cross-sectional area of the sediment in the drain.

WIDTH (m) $\boxed{0.29}$ x DEPTH (m) $\boxed{0.029}$ = CROSS-SEC AREA (m²) $\boxed{0.00841}$

3   Calculate the volume of soil deposited in the drain, where the drain is 10m long.

CROSS-SEC AREA (m²) $\boxed{0.00841}$ x LENGTH (m) $\boxed{10.0}$ = VOLUME DEPOSITED (m³) $\boxed{0.0841}$

4   Convert the total volume to a volume per square metre of catchment.

VOLUME DEPOSITED (m³) $\boxed{0.0841}$ ÷ CONTRIBUTING AREA (m²) $\boxed{50}$ = SOIL LOSS $\boxed{0.001682}$ (m³/m²)

5   Convert the volume per square metre to tonnes per hectare.

SOIL LOSS (m³/m²) $\boxed{0.001682}$ x CONVERSION TO t/ha $\boxed{13,000}$ = SOIL LOSS (t/ha) $\boxed{22}$

Enrichment Ratio

What is it?

Enrichment is the process whereby soil erosion by water tends selectively to affect the finer, more fertile, fraction of the soil, leaving behind coarser, less fertile fractions in the field. Enrichment effectively means that soil material eroded furthest has the highest quality, while soil remaining in the field deteriorates faster because the remaining soil gets progressively less fertile. The enrichment ratio is, therefore, a measure of the proportional enrichment of eroded (and deposited, for example, in drains) materials when compared with the original soil from which they were eroded. It is normally assessed by measuring the quantity of nutrients found in the eroded sediment, compared with the quantity in the topsoil from the field which is being eroded. However, for the purposes of quick field assessment, the proportions of finer soil particles can be used as a proxy measure, as these are closely related to

nutrient levels and in themselves are also good variables for assessment of enrichment. The identification of the occurrence of processes leading to the production of enriched sediment is usually from downslope deposits of sediment. In Plate 28, a sediment fan has been formed by the deposition of fine soil particles carried in runoff from the field. The fan contains particles finer than those that remain on the field, although the very finest will have been carried further downslope.

The enrichment ratio is, therefore, unlike any of the previous measures of land degradation in that it does not give an absolute figure of soil loss. Instead, it assesses the potential seriousness of erosion in accelerating deterioration in soil quality – the higher the enrichment, the more fertility is being lost per unit quantity of erosion. In practice, the enrichment ratio may be used to convert previous field measures such as sediment in drains into absolute total losses of soil.

How does it occur?

Wind and water erosion selectively remove the finer soil particles and lighter organic matter, both of which contain relatively higher levels of nutrients than mineral soils. Thus, when these soil particles are finally deposited lower down in the field, in drains, local reservoirs or eventually the sea, they enrich the location in which they settle. The removal of fines in this way is a natural process, apparent under natural vegetation.

Where does it occur?

Enrichment of sediment occurs almost everywhere. The exact level of enrichment ratio varies from storm to storm, crop to crop, and according to prior history of erosion. Ratios tend to be highest on poorer soils and those with low clay contents. They are also highest at the beginning of the season and immediately after soil disturbance, when there is abundance of fine particles at the surface.

How can it be measured?

Measurement of the enrichment ratio requires comparison of the soil that has been enriched as a result of deposition with the soil from which the deposited material has been eroded. Equal quantities of soil should be taken from the eroded and the depositional locations. By visual observation in the palm of the hand, the proportion of coarse material to fine materials in each sample should be estimated. This should be repeated a number of times. The average percentage of fine materials in both the enriched soil and the eroded soil should then be calculated. The enrichment ratio is the ratio comparing the percentage of fine particles in the enriched soil to the percentage of fine particles in the eroded soil. For example, a single intense storm on a newly tilled soil can give an enrichment ratio of 10:1, which is also simply described as 10.

Potential for error

1 The technique for assessing the enrichment ratio requires considerable field experience, because estimation of proportions of soil particle sizes is difficult. The novice field assessor is best advised to accompany an experienced person.

2 As the selective removal of fines is a natural process, care must be exercised to ensure that the observed trends relate to the land management practices and not to features inherited from prior conditions. For example, ant hills, termite mounds and earthworm casts often contain higher proportions of finer material than the topsoil. Because erosion of these structures may result in the redistribution of this finer material downslope, any observed increase in fines may have little to do with existing land management practices.

3 Estimates undertaken solely by visual inspection of fine particles are very approximate. If possible, laboratory determination of macronutrient (total N, P or K) content or of organic matter should be done to corroborate findings. This is particularly the case for clayey materials.

4 The enrichment ratio can be understated where not all the eroded material is deposited in the site where the enriched soil is identified. The finest particles may have been carried away completely from the site.

5 Understatement of the seriousness of erosion may also occur where deposition from upslope occurs on the eroded soil, thus masking the full extent of finer materials lost.

6 Similarly, the enrichment ratio may be overstated where run-on to the site from further upslope increases the level of fine particles in runoff, thus contributing to the enriched soil.

Worked Example

Field Form: Enrichment Ratio

Site: Farm of Mr Nzirwehi, Bushwere, Uganda.
Date: 13 February 2001

Measurement number	% of fine particles in eroded soil: ie soil remaining in-field	% of fine particles in enriched soil: ie soil caught downslope and deposited
1	20	28
2	25	25
3	15	30
4	22	30
5	20	35
6	20	35
7	22	35
8	19	25
9	20	30
10	20	28
11	18	28
12	20	32
13	18	30
14	22	32
15	22	28
16	20	28
17	18	26
18	20	30
19	20	35
20	19	30
Sum	400.00	600.00
Average	ERODED = 20%	ENRICHED = 30%

Calculations

1   Calculate the ratio of fine materials in the eroded soil to fine materials in the enriched soil.

ENRICHED % $\boxed{30}$ ÷ ERODED % $\boxed{20}$ = ENRICHMENT RATIO $\boxed{1.50}$

Soil Texture and Colour

What are they?

Soil texture is the 'feel' of a soil, constituted by the relative proportions of different types and sizes of particle making up the soil. Soil colour is directly the 'look' of a soil, constituted by the overall hue (based on primary colours), chroma (the strength of the colour) and the degree of greyness (from black to white) of the soil. When soil degradation takes place, both the texture and colour change. These changes can provide opportunities for field assessment of the occurrence and degree of land degradation. Colour changes, especially in recently cultivated land, are often one of the first obvious indicators of land degradation.

How do they occur?

Soil texture and colour are intrinsically functions of the parent material of the soil, as modified by organic material. Dark soils, rich in clays, come from basic rocks such as basalt. Light soils, poor in clays, come from acid rocks such as sandstone. Texture is dependent on the size and shape of particles and, therefore, on the mix of sand, silt and clay making up the soil. Soil texture is important for two reasons. First, particle size and shape influence the likelihood of loss through wind or water erosion. For example, the creation of an armour layer (see 'Armour Layer') results when wind or water action has removed the finer soil particles. Second, soil texture also affects the infiltration rate of water, which in turn influences the amount of surface runoff and the potential for removal of soil particles. In general, the larger the size of particles, the greater the spaces between them and so infiltration occurs more quickly, unless interrupted by other factors. Water stress is a common production constraint, especially in arid and semi-arid areas. Where the moisture holding capacity of the soil is reduced or degraded, the incidence of moisture stress may be more frequent, or occur after lesser periods of dry weather.

How can soil texture be measured?

The generally accepted dimensions of soil particles are set out in Table 4.2

There are standard field techniques for assessing texture, involving feeling the soil in the hand when moist – see any field manual such as that published by FAO (Appendix IV). For land degradation assessment, it will usually be sufficient to categorize texture into:

- Sandy – sand size particles predominate; low intrinsic fertility; easy to degrade (sensitive); fine and medium sands susceptible to wind erosion
- Loamy – balanced proportions of sand, silt and clay, plus usually abundant organic matter; fertile; no major use limitations; difficult to degrade (insensitive)
- Clayey – dominated by clays (either active clays or highly weathered stable clays); susceptible to several degradation processes such as waterlogging; high intrinsic fertility; variable sensitivity to degradation

Texture assessment should be accompanied by the implications for degradation as noted above but supplemented by field observations and the specific nature of the soil.

For land degradation assessment through texture, it is important to select the least degraded soil (from, say, a hedgerow, local forest, graveyard) and compare it with a degraded field soil. If water erosion has been prevalent, the loss of organic matter and selective removal of silt and clay will influence texture. Stones, varying from fragments of quartz to large pebbles, are also aspects of texture affected by degradation – see 'Armour Layer' and 'Pedestals' for assessing these.

Table 4.2 *Classification of Soil Particles by Size*

Description	Size	Visibility (to naked eye)
Sand	0.050–2.000mm	Particles are visible
Silt	0.002–0.050mm	Particles are barely visible
Clay	< 0.002mm	Particles are not visible

How can soil colour be measured?

Munsell soil colour charts give a full description and code for soil colours. It is necessary to standardize the moisture level of the soil for the colour determination. In dry conditions the colour is best examined from naturally dry soil, but in humid conditions moist soil is far more practical. Where soils may be observed in either dry or moist form, it is best to record both colours.

For land degradation assessment, it is again necessary to compare colours between undegraded and degraded conditions. Holding samples of the soil from the two conditions is an immediate indicator of soil degradation. The Munsell colour values give a semi-quantitative measure.

At a larger scale in-field, the occurrence of lighter patches in a field is often the result of topsoil loss and exhumation of subsurface, which is naturally sandier and lighter in colour. On terraces, lighter soil usually occurs on the upslope portion where soil has been removed and transported to the lower part of the terrace. On fields with no barriers, patches are more common, especially on slight rises, spur crests and upper slopes. Examination of large-scale air photos can supplement the field observations of these lighter patches.

Potential for error

1   Precise assessments of soil texture can only be accomplished after laboratory analysis. If laboratory services are available, samples should be carefully collected. The results will then be useful not only to calculate selective loss of clays and organic matter, but also in determining enrichment ratio. Field assessments must be kept very broad, because it takes considerable experience to be able to assess, say, sand percentage to nearer than ± 10 per cent.

2   For both soil texture and soil colour, the baseline soil for comparison is crucial. Relatively undisturbed soil is to be preferred, but it is also useful to compare soils from adjacent fields with different histories of land use. Care needs to be taken that the soils are truly of the same intrinsic type: ie that they would have looked and felt exactly the same if land use had not occurred.

3   Soil colour varies greatly with soil depth. Care needs to be exercised that only surface soil is used, so that differences in colour can then be ascribed to soil erosion.

Soil and Plant Rooting Depth

What is it?

Soil depth is simply the vertical depth of soil from the surface down to weathered rock or other impermeable barrier, such as a stone-line or hardpan. Rooting depth describes the depth available to plant roots – for all practical purposes, it is the same as soil depth.

How does it occur?

The depth of soil material above weath-ered rock is a product of climate, which determines the rate of chemical breakdown of rocks, and the type of rock. Some rocks break down more quickly than others. The specific depth at any one site is determined by the balance between natural forces of removal of topsoil (sometimes called geological erosion – occurs at a rate of less than 1t/ha/year) and the formation of new soil in the subsurface. The faster the rate of weathering and the more susceptible the rocks are to breakdown, the deeper the soil. Deep soils are not necessarily more fertile, because they may contain layers of highly weathered and nutrient-deficient clays.

Soil loss, through erosion by wind or water, serves to reduce the topsoil and thus the rooting depth. The other main way that soil depths and plant rooting depths may be reduced is the formation of an impermeable layer induced by land use or agricultural practices. It may be from ploughing when the soil is too wet, which results in a compacted layer below the plough blade (see Plate 29), or it may form because of chemical compaction in and around stonelines. Formation of an impermeable layer is a direct land degradation process. Box 4.6 describes how agricultural practices resulted in the creation of just such an impermeable layer in Malawi.

The rooting zone is the main supplier of nutrients and water for plants. If the rooting depth available to a plant is insufficient to allow that plant to put down sufficient roots, the plant will exhibit less vigorous growth and crop yields are likely to be depressed. The depth of soil required by different plants varies, as does their ability to put down roots. For example, cotton roots cannot penetrate soil with a bulk density greater than $1.8t/m^3$. Wheat requires 750mm of soil depth, or else yields will fall.

Soil and plant rooting depth are, therefore, important indicators of erosion, because they may directly affect production output if depth is limiting. They are variables most often mentioned by farmers. Hence, they are important to assess, and then to relate to observations of plant growth – see Chapter 5.

How can it be measured?

The rooting depth can be easily measured in a number of ways:

1   Using a soil auger: taking a sample of the soil using an auger shows the different horizons that occur in a soil profile. It may be possible to identify any impediment to the rooting depth from a visual inspection of the soil core from an auger (see Plate 30).

2   Digging a hole: by digging a hole in a farmer's field, the whole soil profile can be identified. The depth of topsoil can then be measured down to a clear indicator of a condition that limits root penetration, such as a line of stones, a change in soil colour or a marked increase in clay content. The

---

BOX 4.6 EVIDENCE OF HOE PAN IN MALAWI

A study in Malawi in 1999 identified the existence of hoe pan as an impediment to root development where crops are planted on ridges. Where these ridges are split and reformed each year, the repeated impact and scraping of the hoe results in a compacted layer. The effects of people walking in the furrows between the ridges, for example during weeding, may exacerbate this compaction.

Evidence of this kind of problem can be obtained from examining plant roots. If roots are stunted and are forced to grow horizontally rather than vertically, then further investigation may reveal the existence of a compacted layer beyond which roots cannot penetrate (see Plate 31).

In the case of crops planted on ridges in Malawi, root depth was restricted to 150mm, being the depth of the ridge.

Source: Malcolm Douglas, Consultant, personal correspondence

distribution of plant roots is also indicative of impermeable layers and the effective plant rooting depth. This method is disruptive, and not really appropriate in a field with growing crops. An alternative, or supplementary, approach is to use road or track cuttings. These often reveal the presence of barriers for roots.

3   Using a stick or steel rod: by applying pressure to a stick/rod, it will pass through layers of soil until it meets resistance that prevents the stick being pushed any further into the ground. This approach will not give an accurate measure of the topsoil depth as the pressure exerted each time the stick is pressed into the soil may not be the same, either as a person gets tired, or if different people undertake the exercise. However, the advantage of this method is that a large number of measurements can be taken, and conclusions reached about relative depths of topsoil in a field.

Potential for error

1   Although shallow topsoil depth may imply that land degradation has taken place, unless the measured depths can be compared with previous measures on the same plot (or some other indicator of the topsoil depth – for example if a house with foundations was constructed, or someone buried or a well dug) or to similar plots that have been managed in a different way, it will be difficult to say with certainty how the shallowness can be explained. Some soils are shallower than others even before land degradation, and in some cases barriers to the rooting depth occur naturally and not as a result of any degrading process.

2   The effective plant rooting depth may be controlled by other factors, such as groundwater or very sandy layers with no nutrients. Therefore, visual inspection of depth should include observation of root distribution and possible reasons for lack of roots in any layer.

5

Indicators of Production Constraints

Land degradation concerns farmers in its effect on production. Most responses from land users to changes in soil quality are tied to some aspect of agricultural production: reduced yields; greater difficulty in maintaining yields; more weeds; stones on the surface making ploughing difficult. The farmers' perspective is, therefore, most often articulated through how production is changing and the way in which plants, soil, water supplies and natural vegetation have deteriorated, making production more problematic. It is therefore essential that this handbook reflects the concerns of farmers, because this is the way they most often make their assessments of land degradation.

Farmers are the primary source of information. They decide on the appropriate indicators of production and they determine whether land degradation is serious for their circumstance. They are able to put current production into context in terms of both historical trends and changes in production methods. They have their own ways of observing and describing the evidence of the effect of land degradation on production. PRA techniques (Chapter 3) may yield much valuable data about the extent of land degradation and how it leads to changes in farming practices over time.

Indicators of production constraints do not give a quantitative measure of the extent of land degradation, as is possible with estimation of soil loss. Instead, these indicators identify problems which may have been caused by land degradation. It is possible that other factors (eg drought stress; intensive cropping) result in the identified production constraints. However, these other factors may themselves also be partly related to land degradation. Drought, for example, may not just be a lack of rainfall; it can also be caused by a reduced available soil water capacity that has been induced by loss of organic matter. Similarly, intensive cropping without replenishment of harvested nutrients leads to reduced future yields. This is, in effect, land degradation but through an indirect agent, the excessive removal of soil chemical fertility in the crop (see 'Nutrient Deficiencies' later in this chapter). Thus, while the identification of these production constraints is not conclusive evidence of land degradation, further investigation may well conclude that this is the most likely direct or indirect cause of the problems.

Because production constraints involve observations and data from many sources, experience has shown they are the most difficult for which to find systematic evidence. Therefore, the checklist questions in Table 5.1 have been devised for field use.

Table 5.1 *Field Questions Checklist*

Questions	Go To Section(s)
What medium to long-term trends in yields do farmers report?	Crop Yield
What do yield records on-farm, from local cooperatives or district statistics show?	Crop Yield
What measures do farmers report for counteracting and compensating for yield decline? [eg fertilizer; change crop; new variety; farm more/less land]	Crop Yield
Do you notice differences in crop size and/or vigour within the field?	Crop Yield and Crop Growth
What crop-specific evidence is there of differential growth? [eg height; number of tillers; size of root]	Crop Growth
What immediate explanations are there for these differences? Factors seen by you; and factors reported by farmers?	Crop Yield and Crop Growth
Are soil texture, soil colour, soil depth clearly related to within-field yield differences?	Soil Texture, Colour and Rooting Depth
Do you see markings or colour differences on the leaves of plants? Any differences between young and old leaves?	Crop Growth and Nutrient Deficiencies
Is there any association of these differences with field evidence of land degradation? [eg yellow leaves and shallow/sandy soil]	Nutrient Deficiencies
Are soil texture, soil colour, soil depth clearly related to markings or colour differences on plants?	Soil Texture, Colour and Rooting Depth
What measures have farmers taken to counteract or compensate for these possible nutrient deficiencies? [eg nitrogenous fertilizer]	Nutrient Deficiencies

Crop Yield

Crop yield is dependent, in part, on the underlying productivity of the soil. It is also affected by seed quality, climate, pests, crop diseases and management by the farmer. The assessment of trends in crop yield, in association with farmers, may show that crop yields have fallen which, in turn, may indicate that land degradation has taken place.

While falling crop yields can be indicative of land degradation, this is not the only possible explanation of decreasing yields – for example, the yields of perennial crops may fall as they get older. Even if yields are increasing, land degradation may also be occurring, but its effects may be masked by the management practices adopted by the farmer, such as increased amounts of fertilizer (see Box 2.4). Indeed, this masking of land degradation by greater and greater use of inputs is considered by some to be the most serious consequence of land degradation, indicating that future yields will crash when farmers are no longer able to afford the inputs. Some of these issues are considered further in Chapters 7 and 8.

An historic comparison of yields can provide useful information about changes in production. By accessing records of past crop yields from farm records, local cooperatives, marketing boards or official government statistics, a good idea of medium- to long-term trends can be

Plate 1 *Ploughing with Oxen, Tanzania*

Plate 2 *Stony Soil Surface*

Plate 3 *Farmer Showing Researchers Evidence of Erosion in his Field, Bolivia*

Plate 4 *Eroded Wastelands in Rajasthan, India*

Plate 5 *Erosion under Cotton Plants, Ghana*

Plate 6 *Eroded 'Badlands': Sodic Soils, Bolivia*

Plate 7 *Tree Root Exposure as a Result of Soil Loss from Steep Slopes, Sri Lanka*

Plate 8 *Land Cleared Using Fire for Conversion to Agricultural Use, Papua New Guinea*

Plate 9 *Discontinuous Gully in Lesotho*

Plate 10 *Constructing* Ngoro *Pits, Tanzania*

Plate 11 *Researcher in Discussion with a Farmer in his Field*

Plate 12 *Field Visit in the Rain, Sri Lanka*

Plate 13 *Interview with Farmer, Bolivia*

Plate 14 *Focus Group Discussion, Mexico*

Plate 15 *Rills, Lesotho*

Plate 16 *Gully, Bolivia*

Plate 17 *Pedestals with Carrot Seedlings, Sri Lanka*

Plate 18 *Measuring Armour Layer Using a Ruler*

Difference between original soil level when the tree was planted and the soil level at the time of observation

Plate 19 *Tree Root Exposure, Vietnam*

Plate 20 *Aerial Roots of Maize, Brazil*

Plate 21 *The Old Silk Road, Gaoligong Mountains, South-western China*

Plate 22 *Scooped-out Depression Caused by Waterfall Effect*

Recent surface

Current surface

Solution notch –
'red' stain, probably
from iron oxides in
the soil

Plate 23 *Solution Notch*

Plate 24 *Tree Mounds*

Note the difference between the level of the soil where the researcher is standing and on the other side of the hedge

Plate 25 *Build-up of Soil Behind a* Gliricidia *Hedge, Sri Lanka*

Plate 26 *Build-up Behind Tree Trunk*

Plate 27 *Sediment in Furrow, Venezuela*

Plate 28 *Sediment Fan, Sri Lanka*

Plate 29 *Plough Pan, Brazil*

Plate 30 *Using a Soil Auger*

Plate 31 *Evidence of Stunted and Horizontal Root Growth in* Acacia mangium

Plate 32 *Sri Lankan Farmer Demonstrating Crop Yield by Making a Clay Model*

Plate 33 *Differential Growth of Radishes In-field, Sri Lanka*

Plate 34 *Differential Maize Growth, Mexico*

Plate 35 *Evidence of Nutrient Deficiencies in Millet Crop*

Plate 36 *Mexican Farmer Showing Difference in Colour between Fertile and Infertile Soils*

Plate 37 *Farmer Planting* Gliricidia *Fence, Sri Lanka*

Plate 38 *Field Showing Poor Maize Growth*

Plate 39 *Maize Planted Up and Down Slope*

Plate 40 *Paddy Fields, Sri Lanka*

gained. Then, putting those records along-side statistics on fertilizer use, introduction of new varieties and other production-enhancing factors, a qualitative view may be gained of how far land degradation may have affected production. Often, however, farmers change their production and liveli-hood practices in response to land degradation. Any one or more of the following explanations and factors should also be considered:

- Change in crop type to one more toler-ant of degraded conditions: eg maize to millet; sorghum to cassava; or annual crops to perennials
- Extensify production onto more marginal hillslopes and poor soils: note that this tends to reduce average yields even faster, and cause further land degradation
- Intensify production on smaller areas by applying manures, irrigation or other inputs: note that this may well reduce overall land degradation
- Land users migrate to towns, or diver-sify sources of income into non-farm activities such as poaching, brewing, charcoal-making or village industry; each of these, in turn, may have land degradation implications.

These coping and adaptation practices in response to land degradation are only amenable to descriptive and non-quantita-tive analysis. The field assessor will want quantitative measures of production constraints. In terms of changed yield, these can be obtained rapidly through participatory techniques directly in the field. Plate 32 gives one example from Sri Lanka where a smallholder has moulded a lump of earth to indicate to the researcher the expected size of radishes from different parts of the field. Within-field differences

in yield are often very significant – the farmer will be well aware of these differ-ences, and the researcher may be able to relate the yield differences to land degrada-tion variables such as soil depth. Root crops, such as carrots, sweet potatoes and beet, are especially amenable to this partic-ipatory technique. Farmers are also often happy to draw the size of their individual root crops onto paper. The researcher may then purchase an equivalent size of crop from the market, weigh it and multiply by the number of plants in a fixed area to get accurate yield assessments.

Two further examples arose in Uganda during the final validation workshop for this handbook. Researchers wanted to know how banana yields varied in differ-ent parts of a farm in relation to measured soil losses and they developed a technique with the farmer, correlating stem circum-ference with bunch size (see Box 3.2). In the second case the researchers were told by the farmer that the higher his maize plants grew, the longer were the cobs and the better the yield. Jointly collecting data from a representative sample of plants, a relationship was derived between the vari-ables of plant height and yield per plant. This was then used for the fuller survey to assess the impact of land degradation on yield.

Other practical yield assessment tech-niques that have been used in the field are listed in Table 5.2 and should be consid-ered for application in appropriate situations. A word of warning, however: information on yield will depend on human recall. The limitations of memory must be recognized – it provides a personal history and interpretation rather than factual evidence. Yet, it is the farmer-perspective that it is vital to obtain, rather than absolute quantitative yield figures.

Table 5.2 *Techniques for Assessing Yield*

Field-based Yield Assessment	Relevant situations and warnings
Relative diameter of growing crops in relation to land degradation indicators, such as depth of topsoil, organic carbon content or slope	This is useful for vegetables, planted on same date but in different parts of the field. Lettuces or cabbages have significantly different diameters according to soil quality – these measures are a good proxy for yield, especially if the farmer can show what size they are expected to reach at harvest
Relative height of growing crops (as above)	Height is a good proxy of yield for other crops, such as maize. But note that height is very specific to crop variety, and so relative measures can only be used for the same variety
Number of tillers on individual cereal plants, such as wheat, barley and oats	For many cereals, the number of tillers is directly related to yield, because each tiller has a seed head. So, a count of tillers is a useful proxy for yield. Again, the farmer can help by indicating size of expected seed head
Plant population per square metre	Where germination is poor due to land degradation, plant population in degraded versus less degraded parts of the field is a useful proxy. This has been used with cereals, especially where soil crusting by rain drop impact has affected germination
Direct farmer assessments of bags of marketable yield per field from growing crop	From experience farmers will usually be able to estimate the number of bags of crop yield. Comparison of farmers' estimates between fields is especially useful

Crop Growth Characteristics

Several of the yield assessments use crop growth as a proxy for yield. However, crop growth characteristics by themselves are  one of the most common indicators of plant vigour described by farmers. In so far as crop growth is related to land degradation, observations and simple relative measurements are very useful in obtaining a farmer-perspective. Crop growth characteristics depend on the seed itself, the agronomic practices followed by the land user, the soil and the climate. Within fields it may be possible to identify differential crop growth. Often this is very clearly visible (see Plate 33). In other cases, the crop needs closer inspection and field measurment to ascertain the differential growth. The question that must be asked is 'what has caused this difference in growth pattern throughout the field?'

While it may seem that the cause of differential plant growth is self-evident, it is worthwhile taking some time to map the incidence of the differential growth, and then to plot the possible causation factors. The mapping of the growth is most easily achieved by dividing the field into a grid and recording the relative vigour of the plants in each square. In determining the reasons for differential growth, it is important to eliminate as many explanations as possible. The example in Plate 34 from highland Mexico was identified only with the help of the farmer as relating to the local extraction and application of gypsum. The farmer had had labour sufficient for only part of the field. A checklist of questions to help identify the reasons for differential crop growth might include the following:

Crop factors

- Are all the crops in the field the same variety? Very often land users will elect to plant a mix of high yielding (for sale) and lesser yielding (for home consumption and taste preference) varieties that will, nevertheless, produce some yield even if the growing season is particularly dry or wet, or particularly hot or cold.
- Were all the plants in the field sown or introduced at the same time?
- Are the row distances constant throughout the field, or are crops planted more densely in some parts of the field than others?
- Do plants in one part of the field show signs of pest infestation/consumption that are not on plants elsewhere in the field?

- Have animals been grazing along the field boundaries, resulting in reduced crop density and vigour?
- Has one part of the field had a different treatment applied to it?

Land degradation factors

- Are parts of the field more exposed to wind than the rest?
- Are parts of the field more sloping than others?
- Have conservation or tillage practices introduced in-field differences in soil depth or accumulations of fertile sediment?
- Are there accumulations of soil behind barriers, such as boundary walls and hedges?
- Has farming practice caused 'plough erosion', ie the progressive removal of soil downslope by hand or with the plough?
- Are any parts of the field inherently more fertile than others (eg old stream beds)?

Knowledge of the common characteristics of locally planted varieties is extremely useful in determining how a crop that is uniformly productive on a particular plot compares with the same crop planted elsewhere in the locality. Comparisons with fields of the same crop planted nearby may suggest that different management practices have been followed.

Nutrient Deficiencies

Nutrient deficiencies are one of the most common ways in which land degradation affects production. Hence, it is essential for the field assessor to be aware of the evidence of such deficiencies in growing plants. Plate 35 from India illustrates the field identification of nutrient deficiencies in millet plants. In most cases, by the time nutrient deficiencies are evidenced by abnormalities in the visual presentation of a plant, it is already too late to correct the deficiency in time to affect current yields. Nevertheless, if future productivity is to be maintained or increased, it is important to identify, as far as is possible, the cause of the abnormalities. As will be discussed below, this is not a straightforward task.

Different crops require different levels of nutrition. This means that some species may be more susceptible to particular deficiencies than others. Land degradation can, therefore, affect some crops and leave others untouched. So, as with yields and crop growth characteristics, the effect of deficiencies of nutrients, resulting from land degradation, is both crop-specific and soil-specific. This is why local people may respond to nutrient deficiencies by applying fertilizers and manure or changing to a less demanding crop. These responses are themselves also good evidence of nutrient deficiencies, which can be gained from local people and their explanations of why they have changed practice.

Nutrient deficiencies are caused by more than just removal in the processes of soil degradation. The principal cause (up to 100kg nitrogen or more, in intensive cropping) comes from removal in harvested crops and insufficient replenishment through manures or fertilizer. Excess removal through harvesting, although unrelated to soil erosion, is still a factor of land degradation. Thus, in determining the cause of nutrient deficiencies, the field assessor must make careful judgement, tying field evidence with other aspects of farming practice and local knowledge.

Many commentators argue that visual symptoms are not sufficient indicators on which to base conclusions about nutrient deficiencies or toxicities. The main reasons why visual symptoms alone are insufficient for determining the existence of nutrient deficiencies and their link to land degradation are:

1   Different plants respond in different ways to nutrient deficiencies. For example, root crops demand over twice the levels of phosphorus needed by cereals or beans.
2   Deficiencies (or toxicities or other degradation factors) of different nutrients may exhibit the same visual symptom. For example, yellowing of bean leaves can indicate lack of nitrogen, waterlogging or even salinity. In maize, the accumulation of purple, red and yellow pigments in the leaves may indicate nitrogen deficiency, an insufficient supply of phosphorus, low soil temperature or insect damage to the roots.
3   Disease, insect and herbicide damage may induce visual symptoms similar to those caused by micronutrient deficiencies. For example, in alfalfa it is easy to confuse leaf-hopper damage with evidence of boron deficiency.

Notwithstanding these valid objections to the use of visual observations, their judicious use can provide valuable insights into the constraints in particular cropping systems.

Indicative Conditions for Nutrient Deficiencies

Certain soil types, or soil uses, may be more likely to display nutrient deficiencies than others. The combination of particular soil conditions with visual indicators of nutrient deficiencies makes the conclusions drawn from the latter more robust. In Table 5.3 some of the conditions that can lead to nutrient deficiencies and toxicities are noted. These are not the only situations in which deficiencies or toxicities may occur. Land management practices also have a significant impact on the potential for nutrient deficiencies/abnormalities.

Table 5.3 *Nutrient Deficiencies and Toxicities: Generalized Symptoms and Circumstances*

Essential nutrient	Deficiency/toxicity symptoms	Typical conditions
Nitrogen (N)	Leaves (first older ones) turn yellow/brown, plants are spindly, lack vigour and may be dwarfed	Sandy soils under high rainfall conditions and soils low in organic matter, where leaching occurs
Phosphorus (P)	Not easily detected from appearance. Where deficiency is severe, plant will be stunted, the leaves will take on a purplish tint and the stem will be reddish in colour	Acid soils rich in iron and aluminium oxides (ie red tropical soils)
Potassium (K)	Yellow/brown spots appear on older leaves and/or necrosis of edges	More frequent on light soils (as K is concentrated in the clay fraction of soils)
Sulphur (S)	Leaves are stunted, with uniform chlorosis	Most frequent in light, sandy soils
Calcium (Ca)	Roots are usually affected first – growth is impaired and rotting often occurs. In vegetative growth, deficiency may show in distorted leaves, brown scorching or spotting on foliage or bitter fruit (eg apple) or blossom-end rot (eg tomato)	Acid soils, or alkali or saline soils containing high proportions of sodium
Magnesium (Mg)	Interveinal chlorosis, first on older leaves	Acid, sandy soils in areas with moderate to high rainfall. Often occurs in conjunction with Ca deficiency
Iron (Fe)	Chlorosis of younger leaves	Calcareous soils, poorly drained and with high pH. (In neutral and alkaline soils P may prevent the absorption of Fe)
Manganese (Mn)	Chlorosis of younger leaves	Badly drained soils, over-liming or deep ploughing of calcareous soils can lead to Mn deficiency, as can the presence of high levels of Mg. The combination of high pH values (> 6.5) and high levels of organic matter can immobilize soil Mn
Zinc (Zn)	Symptoms vary with plant type – in cereals, young plants display purpling, whereas in broad-leaved plants symptoms include interveinal chlorosis, reduced leaf size and sparse foliage	Soils with high pH. Available Zn is reduced by the application of lime or phosphates
Copper (Cu)	Chlorosis of the tips of the youngest leaves and die-back of growing points	Peat soils, or leached sandy or acid soils
Boron (B)	In crops other than cereals, the apical growing point on the main stem dies and lateral buds fail to develop shoots	Sandy soils, dry conditions and liming can result in B deficiency
Molybdenum (Mo)	Marginal scorching and cupping of leaves. Wilting is common in Brassicas	Acid soils or soils with high pH. Mo deficiency can lead to N-deficiency as nitrate requires adequate supplies of Mo for metabolism. Mo availability can inhibit the uptake of Cu

Table 5.3 *continued*

Essential nutrient	Deficiency/toxicity symptoms	Typical conditions
Chlorine (Cl)	Wilting of leaves	Well-drained, sandy soils
Sulphur Toxicity	Failure of plants to grow. Large bare patches in fields	Build-up of sulphates as a result of irrigation. Drainage of acid sulphate soils
Manganese Toxicity	Brown spots and uneven chlorophyll in older leaves	Soils with pH of < 5.0 (for susceptible species)
Copper Toxicity	Chlorosis of leaves and restricted root growth	Soils with low pH
Boron Toxicity	Progressive necrosis of the leaves, starting from the tips and/or margins	Soils with low pH
Aluminium (Al) Toxicity	Plants die after early growth	Acid mineral soils, aggravated by low P status
Chlorine Toxicity	Burning of leaf tips, bronzing and premature yellowing of leaves	Associated with irrigation using water containing chloride

Identification of Nutrient Deficiencies

Observation of abnormalities in plants is a complicated and skilled task. Since nutrient deficiencies may be manifested in different ways depending on the crop in which they occur, particular criteria will be crop-specific. As an example, the visual indicators of nutrient deficiencies in three crops are set out in Table 5.4.

Measuring Nutrient Levels

In most cases, plant analysis is carried out in laboratories. However, it is possible to test the levels of various nutrients in the field. There are two simple ways to do this:

1 **Using manufactured strips:** Some manufacturers produce strips that react with plant sap. The sap can be squeezed from the leaf-stalk onto the strip, which will change colour depending on the level of the nutrient being measured. (These strips may require refrigeration before use, and so may not be suitable for use in remote fields.)

2 **Using filter paper and colour reagents:** Sap tests for particular nutrients can be performed using colour-developing reagents. Sap extracted from leaves is smeared onto filter paper to which the reagent is applied. The resulting colour is then compared to a standard colour chart to determine the level of the nutrient being measured. Different reagents are used, depending on the nutrient being measured.

Nutrient depletion can be a cause of further land degradation since nutrient-poor soil produces less biomass, both above and below ground, which gives poorer protection to the soil from erosion. (Above ground the plant canopy protects the soil surface from the impact of rain-drops, while below ground the roots bind the soil, and when dead add humus, promoting aggregation.)

Table 5.4 *Examples of Deficiencies in Three Crops*

	Maize	Beans	Cabbage
General	High N requirement and sensitive to low phosphate supply. Relatively sensitive to water stress	Tolerant to a wide range of conditions, but only high yielding with high N	Demanding of N, P and K. Moderately sensitive to water stress
Nitrogen	Reduced vigour; leaves a pale green or yellowish colour	Plants are small, leaves are pale green and older leaves turn yellow. Few flowers are produced	Young leaves pale green, older leaves are orange, red or purple. Severe deficiency renders the crop useless
Phosphorus	Stunted growth, delayed ripening and purplish leaf colour, especially during early growth	Stems are dwarfed and thin, leaves lack lustre. Early defoliation occurs, starting at base of shoot	Leaves are dull green with purplish tinge, margins die
Potassium	Small whitish-yellow spots on leaves. Poor root system, plants are weak and may be blown down	Chlorosis of leaves, with necrotic brown patches at margins between veins	Leaves are bluish-green. Leaf margins may show scorching and tips of older leaves may die
Sulphur	Somewhat similar to N-deficiency. Plants short and spindly. Younger leaves pale beige to straw in colour	Stunted growth, yellowing leaves. Delayed flowering and development of beans. Reduced nodulation on roots	Smaller plants, with yellowing leaves
Calcium	Poor germination and stunted growth	Growth is stunted and growing point may die. In severe cases plants turn black and die	Leaves rolled up at margins, necrosis of rims and death of growing point
Magnesium	Whitish or yellow striping between the leaf veins, followed by necrosis	Older leaves show interveinal reddish-brown mottling	Interveinal chlorosis and puckering of older leaves
Iron	Alternate rows of green and white on leaves	At early stage, patternless paling in leaf colour; later stage, yellowing of leaf similar to N-deficiency	Whitish streaks on leaves. Veins unaffected at first, but larger veins eventually turn yellow
Manganese	Yellow and green striping along the length of the leaf	Chlorosis, initially of young leaves, followed by necrotic spots in inter-veinal areas. Leaves will fall off and plants eventually die	Leaves are of smaller size and exhibit yellow mottling between veins
Zinc	Chlorotic fading of the leaves, with broad whitish areas	Leaves and flower buds are shed	General chlorosis and failure of plant to develop a head
Copper	Leaves become chlorotic and the tips wither	Roots and leaves stunted, especially on organic soils	Leaves chlorotic, heads fail to form, growth stunted

Table 5.4 *continued*

	Maize	Beans	Cabbage
Boron	New leaves show transparent stripes. Growing points die and ears may not develop	Leaves turn yellow and then brown. No flowers or pods are produced	Leaves are distorted, brittle, mottled along margins and wilted
Molybdenum	Not common by itself, but indicators include scorched patches on leaves	Leaves are smaller, pale in colour with interveinal mottling developing into brown scorched areas	Older leaves become mottled, scorched and cupped. Margins are irregular and heart formation is poor
Chlorine	Plants short with poorly developed stubby roots	Cl essential for the symbiotic fixation of N in legumes. No nodulation and stunted growth	Stunted roots with excessive branching and poor wilted top growth

Soil Variables Related to Production: Texture, Colour and Depth

Three of the most frequently observed soil indicators that farmers relate to production constraints are soil texture, colour and rooting depth. These are easily observable, and it is clear in the minds of most land users how they relate to changes in plant growth. The methods of using them as indicators of soil loss have already been covered in Chapter 4. Here, they are related to observations on production, and their use as supporting measures/indicators for assessing the effects of land degradation on production.

As described in Chapter 4, soil texture, colour and depth are intrinsically a function of the parent material of the soil and rate of weathering. The three variables directly link to production through the biophysical processes of plant growth in supplying nutrients and water and providing a medium conducive to plant growth. By assessing one or more of these variables simply in the field, the field assessor has an excellent check on how and why production has changed consequent on the processes of land degradation.

Of particular importance in gaining a farmer-perspective is the fact that these three variables of soil – colour, texture and depth – are routinely used by farmers to assess the productive potential of their fields (see Plate 36). Where farmers bring up lighter coloured subsoil with the plough or hoe, they will always recognize that yields will be less because of the thin topsoil and lack of rooting volume. Texture, also, especially dark clays and organic matter, is a common indicator. Farmers will relate how putting in manure or growing cover crops changes the 'feel' of the soil. They will also note the occurrence of coarse particles and stones as indicating poor productive potential. Soil depth is closely related to soil colour, but thin soils will be noted as particularly limiting for some crops, especially root crops and cereals that are demanding of water, such as maize.

These observations by farmers present useful opportunities to the field assessor. Soil variables enable a check to be made on other indicators of production change such as yield statistics – this is an example of triangulation (see Chapters 1 and 6). Local soil types can be colour-calibrated in order to assess loss of potassium and phosphorus from fields. Accumulations of finer material behind barriers can be assessed in the field for their organic matter content, which is then related to observations of plant growth.

This handbook recommends, then, that along with observations on growing crops and nutrient deficiencies, the following soil variables are noted to provide not only an explanation of production constraints induced by land degradation but also corroborative evidence (see Chapter 4 for more detail):

- **Soil texture:** feel between the fingers – look especially at finer textures and organic matter between a degraded site and another with better production. A baseline measure should be taken from natural forest or grass pasture.
- **Soil colour:** Munsell soil colour chart – look especially for lighter colour patches within fields, and overall lightness of fields compared with a baseline of natural forest or grass.
- **Soil depth:** using a sharp stick to probe to impenetrable subsoil, or digging a pit (see Box 5.1) – look especially for the coincidence of shallow depths and smaller crops and poor germination.

BOX 5.1 EXTENDED SPADE DIAGNOSIS

This is an evaluation technique developed in the 1930s to assess the effects of management practices on soil structure. It simply involves removing a spade-size sod, examining the exposed root structures and undertaking field tests for shear strength and aggregate stability. (For more details see Herweg et al, 1999; see Appendix IV.)

Source: Karl Herweg, CDE, personal correspondence

Facing Problems with Production Indicators?

Production constraints imposed by land degradation are many and varied. This chapter has highlighted how their identification can lead the field assessor to significant insights into the complex relationship between the land user and the quality of the land. Interpretation, however, can often be difficult. For example, without further examination and experiment, nutrient deficiencies can be elusive to detect. Impacts on production are also problematic because there are so many other factors that can also affect yields. It would be easy to become mired in the problems of accurate detection and lose sight of the ultimate purpose of these field observations.

First, degradation-induced production constraints are the main way that land users express how land degradation processes affect them. So, searching for these indicators is an immediate way for the field assessor to target the real concerns of local people and enter into a useful dialogue. Second, the understanding of production constraints is an ideal way of bringing together the biophysical indicators that assess site-specific processes and impacts that affect the land user. Therefore, throughout this chapter a link has been drawn between evidence of land degradation and production problems. For example, accumulations of sediment give evidence of a production opportunity in useful planting sites for demanding crops, while at the same time indicating soil loss from the slopes above and possibly inadequate soil depth to produce a yield. In the next chapter (6), ways of combining indicators to get added understanding are explained. Third, the ultimate purpose of all these field observations is to help to advance the cause of land rehabilitation and conservation. Chapter 7 starts to address this objective by ensuring that the field assessor understands all the 'actors' involved, and how some lose through land degradation while others gain. Then, in Chapter 8, this handbook directly targets

simple means of assessing the benefits of conservation from the farmers' perspective. So, production constraints can prove difficult to identify in the field, at least at first. However, they provide an important part of the big picture of how land degradation affects society. It is well worthwhile delving deeply into the several indicators, using the evidence from the farmer and from one's own eyes. When used alongside other indicators of land degradation (see Chapter 6, 'Combining Indicators'), production constraints lead the field assessor directly to potential solutions that the farmer is likely to appreciate.

6

Combining Indicators

Introduction

Single indicators give singular items of evidence for land degradation or its impact. They are susceptible to error, misinterpretation and chance. Especially in the case of field assessment where many of the measurements can only be described as 'rough and ready', the use of only one indicator – say, a tree mound – to conclude definitively that land degradation has occurred is problematic. It renders the field assessor open to criticism that much is being made of little. Therefore, this chapter addresses how, by combining indicators, more robust conclusions can be entertained, even to the extent that quite different types of measure may be placed alongside each other to obtain a fuller understanding of whether land degradation is happening.

This handbook has throughout promoted the use of two or more indicators in combination, preferably with the active input of farmer experience. Just as a three-stranded rope is far stronger than the sum of the strengths of its individual strands, so is an assessment of land degradation based on the combination of indicators that all tend towards the same conclusion. While each indicator has its own attributes and applications, several indicators combined can piece together a far more comprehensive and consistent picture. Similarly, if indicators disagree in general trends, then the field assessor is led to further investigation to resolve the disparities. Disagreement on what the indicator suggests is one of the most powerful ways of picking up the difference in perspectives of the land user and the field professional.

Three particular areas of combining indicators are highlighted here:

1   Combinations to show both the process and likely cause of land degradation through time
2   Combinations to provide corroborating evidence and a consistent view of land degradation
3   Methods to bring individual indicators together for comparative and overall assessment, including how to search for a suite of indicators and how to develop a semi-quantitative procedure for getting an overall picture

However, before considering how indicator combinations can be constructed, why are they really necessary in all but the simplest of situations?

Why Single Indicators are Often Insufficient

An example is used here to illustrate how single indicators, especially when used with little reference to the farmer, may give erroneous conclusions.

Accumulations of sediment behind barriers are a useful indication that soil movement has taken place in the field and that, if it were not for the barrier, soil would have been irrecoverably lost. A typical example is shown in Figure 6.1, where the sediment trapped (shaded) by the constructed riser of the bench terrace can be measured to give an assessment of the minimum amount of soil that has been lost from the bench. The assumption here is that the material trapped has been eroded from the bench because of the land use and slope of the bench. Is this necessarily so?

In Sri Lanka, these types of bench terrace are common in the steep hill lands where a living plant is used to support the riser and maintain the benches. (See Chapter 8 for a fuller case study of this type of conservation.) It is instructive to follow through with farmers how they construct the benches and what they think of the soil that is trapped behind the riser. The risers themselves usually consist of planted lines of a fast growing leguminous tree, *Gliricidia sepium*. A farmer will culti-vate the field and then plant a new line of *Gliricidia* sticks across the slope (see Plate 37). At this stage there are no benches. Progressively, over a few seasons, the farmer then:

- manages the *Gliricidia* sticks so that they 'strike' and commence growing into trees;
- places sticks along the line to form a more permanent barrier; weeds are also placed here, as it is a convenient close place and does not interfere with crop activities between the lines of sticks;
- scrapes soil downslope with the hoe to cover the weed-fill and form benches; some soil is washed down naturally by rainwash, but most is what is called 'plough erosion', that is, soil moved down by the action of cultivation.

Therefore, we have a situation where erosion has been 'encouraged' by the farmer. Is this land degradation? The field assessor must decide. On the one hand, it is actual physical removal of soil from one place to another. On the other hand, it has been done deliberately. It is useful to view the whole process through the eyes of the farmer because:

- here in the *Gliricidia* hedge is the creation of a useful site to get rid of weeds and other 'rubbish' – indeed, the fence line in the local language is called 'rubbish-things fence';
- after a few seasons when the benches are formed, the soil close to the fence is relatively rich in organic matter as well as being deep; hence the more valuable and demanding crops are planted here;

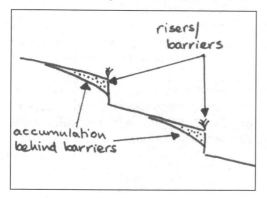

Figure 6.1 *Sketch of Bench Terraces*

- the farmer harvests poles of *Gliricidia* for sale or use as bean-poles, while the leaves are left on the surface soil for a nitrogen-rich mulch;
- after six or seven years, when the *Gliricidia* starts to lose its vigour, the farmer uproots the fence line, and plants high value crops in the accumulated rich soil;
- at the same time, a new *Gliricidia* fence line is constructed midway across the old bench ... and so the process continues.

Only with the farmer fully participating can this story of soil movement and farm production be told. So, again, is this land degradation? First, there is positive encouragement by the farmer for soil to move to fill in the upslope side of the fence. Second, there are interesting management and production opportunities opened up by the accumulation of soil. Third, the farmer sees the accumulation as a longer-term production opportunity, while a new fence line in another part of the field is established. There has been a real and measurable movement of soil. But humans have done most of it for very specific reasons related to their livelihoods. The soil movement will have contributed to the deterioration of part of the slope for some six to seven years. But the farmers gain sufficient capital assets to implement a further cycle of soil restoration with the new fence lines, while fully utilizing the 'eroded' soil for their benefit.

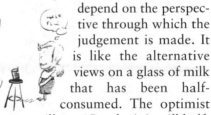

The answer to the question of whether this is land degradation must, therefore, depend on the perspective through which the judgement is made. It is like the alternative views on a glass of milk that has been half-consumed. The optimist will say, 'Good – it is still half-full'; the pessimist will complain, 'It is half-empty'. The optimist will be the farmer. In field surveys in Sri Lanka, not a single farmer equated the soil movement with soil erosion – they saw it as part of the natural production cycle on steep hill slopes. The pessimist is the professional – soil is moving downslope, and what is worse is that farmers are even accelerating the process with their cultivation techniques!

The single indicator, with little reference to the farmer, could in this instance (and many others) present a simplistic and erroneous – from the farmer-perspective – understanding of the status of land degradation. The fuller picture is only available by, for example, examining plant growth on the eroded soil; the use of the *Gliricidia* branches and leaves; and, crucially, by observing what is done and talking with the farmer as to why things are done in this way. The single biophysical indicator needs to be supplemented by all these other observations before land degradation can even be considered as having occurred.

Assessment of Both Process and Cause

The example above has already illustrated how process and cause can be discerned in a complex field system of bench terraces. The field observation indicated that there had been a 'process' of erosion; the further enquiry found the 'cause', deliberate ploughing and entrapment of sediment by the farmer. Conclusions just as powerful may also be drawn in simpler situations where two indicators essentially agree but one is a measure of the process and the other a measure of how the process comes to have an impact on production.

Take the case of a flat, cultivated field where observation and measurement of armour layer has shown that active current erosion is taking place under an extremely poor cover of maize. Plate 38 illustrates one such case being demonstrated by a District Agricultural Officer to his staff and a local farmer in Tanzania. The coarse stones accumulating on the surface are evidence, not only of total erosion, but also that there is substantial selective removal of fine particles. However, the impact of this selective removal and how the erosion is causing a reduction in plant vigour (and presumably also of yield) have yet to be discovered.

Examine the growing maize plants and other field indicators for:

- nutrient deficiency symptoms – to discover if there is a causative effect through plant nutrient limitations: *yellowing, chlorotic leaves of the maize*;
- differential crop growth characteristics between eroded and uneroded conditions to determine if it is the erosion that is reducing yield, and by how much: *a nearby site cultivated for only*

a few years has double the plant density and much larger plants;

- any sediments entrapped in field ditches, or hollows, for evidence of the degree of enrichment: *some coarse sands in a field ditch in the middle of the field; and some very fine clays and rich humus in a puddle at the bottom of the field.*

All add to the understanding of how the obvious process of erosion under a poor standing crop affects current and future production from the field. In this case, the indicators are all in agreement in the sense that they all point to a consistent process (erosion-induced loss in soil productivity) and cause (selective removal of organic matter and clays and consequent nitrogen deficiencies for the maize).

Piecing together the separate strands of field evidence is one of the most exciting aspects of field assessment of land degradation, because it enables far more to be learned than with classical reductionist methods, where only the knowledge of a process may be gained. Here, an interesting interweaving of process, cause and effect may be gained, provided that the field observer is alert to the signs and is willing to put together evidence from a variety of sources.

Triangulation: Gaining a Robust View of Land Degradation

Mention has already been made in Chapter 1 that farmer-perspective field assessments of land degradation may be criticized by some as being less reliable than standard measurement. The principal ways to overcome any possible lack of precision are (a) to take as many individual assessments as possible, and (b) to examine the general trends of several different types of measure to see if they are in agreement. This second means is known as **triangulation**, the gaining of a consensus view of overall trends from different types of technique, source of information and investigators (see Box 6.1).

BOX 6.1 TRIANGULATION

Triangulation is used to check evidence gained from different sources and to investigate further the reasons behind inconsistencies. The word is derived from the survey techniques of mapping out an area by tracing a network of triangles from a base-line point. If the surveyor failed to return to the original point, then it would be known that a mistake had been made which needed correcting.

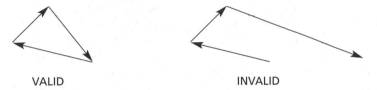

VALID   INVALID

Essentially, then, triangulation is the cross-checking of information by the following means:

1   Multiple techniques: the use of a variety of techniques alongside each other to see if the trend in results is broadly consistent. (Note that different soil loss measures give different absolute results – so it would be wrong to expect them to be exactly the same.)
2   Multiple sources of information: similarly, the use of information from different sources, such as different farms that have broadly similar land uses or different statistical sources for yields.
3   Multiple investigators: cross-checking is aided by two or more field assessors independently making their assessments and then meeting to compare results.

Even more robust is the triangulation from different techniques, sources and investigators – see the example in this chapter.

Example of triangulation using nine indicators

Take the example of a degraded catchment that has been largely deforested in order to plant annual crops of maize and beans with no obvious dedicated measures of soil and water conservation. There are some trees and field-plot boundaries. The maize is planted in rows up and down the slope (see Plate 39).

It is now two months since the rains started. A reconnaissance field survey with the farmer has revealed the following, with some preliminary measurements:

•   In the furrows between the rows of maize, rain-wash has concentrated and formed rills within the planted beds of maize. These rills are discontinuous and some contain the remains of organic matter. Closer field inspection shows there is an average length of rill of 4m; cross-sections average 5mm wide by 5mm deep; and the average contributing catchment to each rill is 1m wide (the row width) and 5m long. So each rill has a space volume of $0.01m^3$ per $5m^2$ of field. The organic matter seems to come from grasses and small herbs. The farmer observes that these rills occur every year, and he finds them useful as narrow paths to get into his field for weeding, as well as places in which to put the weeds.

•   The soil has a significant number of coarse quartz fragments some 2–3mm across. Between the rows of growing maize, these fragments provide the capping material for **pedestals**. A sample of pedestals gives a mean height of 2.5mm. The farmer confirms he last weeded with a hoe three weeks previously.

- There are several trees within and around the field that have been left for shade after a hot day's weeding and for their wild fruit. **Tree mounds** are apparent, indicating that the surface of the soil in the field has lowered because presumably topsoil has been washed off since the field was opened for cultivation. According to the farmer this was 20 years ago. The mounds average 150mm in height above the surrounding soil surface, though there is some considerable variation between top (higher – up to 300mm) and bottom (very little) of field.

- While at the downslope end of the field, our observer notes that there are **boundary accumulations** of soil that average 100mm deep against the grass path between this field and the one immediately downslope. Examining the accumulations more closely, a rough calculation indicates an average volume accumulation of $0.01m^3$ per metre length of boundary. Since the field is 10m long, the contributing area is $10m^2$, and the sediment therefore amounts to $10m^3$ per hectare. The farmer interjects at this stage that he only subdivided the field the previous year and sold the downslope part to his neighbour, and so the path has only been there for just over a year.

- Walking then to the middle of the field, the observer notes that the farmer has constructed a small drainage ditch across the slope to protect the lower field from runoff during heavy storms. There is **sediment in the within-field drainage ditch**, amounting on average to $0.001m^3$ per metre length of drain. The sediment is mainly medium to coarse sand – the fines have apparently been washed completely out of the field. Since each metre of drain has a contributing area of $5m^2$, this amounts to $2m^3$ of sediment per hectare. The

farmer tells our observer that he has to dig this drain out each year as it fills up, and redistribute the sediment across the field, or else the drain will not work.

- While at the top of the field, our observer digs a small hole to examine the soil. **Soil depth** is very shallow, averaging only about 250mm, with little differentiation in colour (a light yellow-brown) between subsoil and topsoil. The farmer says he is getting worried about this part of the field and has noticed the soil getting lighter and sandier. When he started cultivating there 20 years ago, it was 500mm deep with 100mm rich topsoil. At this stage, the farmer gets his hoe out and shows the field assessor how he cultivates: standing facing uphill, the farmer progressively brings soil downslope – this is an immediate explanation for the lack of soil depth here at the top of the field.

- Walking into the maize crop with the farmer, the observer notes that some parts of the field seem to be doing well, while other parts have suffered stunted growth. **Within-field variation of crop growth** is significant, with the upper parts generally poorest. Germination rate as evidenced by plant population density, however, seems to be relatively uniform.

- **Maize nutrient deficiencies** are also evident in the leaves of the growing crop. At the top of the plot, plants are stunted and yellow-looking. Towards the lower and middle parts of the field, some of the plants have a purplish colour on new leaves, but those plants growing in the sediment accumulation along the boundary are sturdy, vigorous and deep green in colour.

- Then, finally, our observer walks with the farmer to the lower boundary of the field to see if there is any evidence of land degradation processes outside the immediate field. There, in a hollow, is some fine mud and organic material obviously collected after the last rainstorm from soil that had been completely washed out from the field. Here the **enrichment of sediment in the downstream hollow** can determine the quality of the material that has been entirely lost from the field. The clay and organic matter amount to 100 per cent of the sediment in the hollow, whereas in the field clay is less than 20 per cent. This indicates an approximate enrichment of the eroded sediment by a factor of 5:1.

In this example, the nine different types of measure all indicate that processes of land degradation are operating. They all show different parts or different aspects of degradation processes that have been set in train from when the land was originally opened up for cultivation. So there is a general consistency in trends, but the evidence is complex. Our field assessor can certainly conclude that there has been degradation and it is having a significant (and increasing) impact on crop growth in parts of the field. However, the simple calculations of the absolute levels of soil erosion from pedestals, rills, tree mounds, boundary wall accumulations and sediment in ditches do not agree. This is unsurprising, because they represent different spatial and temporal scales, as well as different parts of the overall process of land degradation. Some measures give a view of the erosion for the last three weeks (pedestals). One shows what has happened since the field was last ploughed two or more months previously (rills); one from the last year (boundary wall accumulation); right up to one which integrates the situation of the field and its land use for the last 20 years (tree mounds). The spatial scales vary from being representative of a single point on the slope (tree mounds) to half the field (sediment in drainage ditch) and the whole field (boundary accumulations), and even the whole slope (enrichment in downslope hollow). So, it is necessary to examine the different items more closely (Table 6.1) and piece together a comprehensive picture.

The comprehensive picture

In the above example, triangulation has provided the field assessor with powerful conclusions that land degradation is currently active. Also, there is a substantial current effect on production and the loss of fine material is potentially serious to future yields. Previous land use probably also saw degradation, but at a lower rate.

The evidence all indicates that the major influence on land degradation has been the opening up of this piece of land to annual arable crops without any form of protection or conservation, other than the field drainage ditch and the new grassed path acting as a lower field boundary. Overall current sheet erosion rates are at least 100t/ha/yr, with possibly another 25 per cent addition to account for rilling. Only a small percentage (*c.* 20 per cent) of this sediment is caught in-field – 10 per cent of coarse sands in the drainage ditch and 10 per cent of more representative fractions of the whole soil against the field boundary. Erosion-induced loss in productivity is also serious, through a large reduction (50 per cent) in plant-rooting volume at the top of the slope, which affects nutrient and water supply to growing plants. The erosion-induced limitation in mid-field is a reduction below critical threshold of available (soluble) phosphorus. At least two-thirds of the field is affected by serious land degradation, while the lower third has gained somewhat. However, still a very large percentage of fine particles and organic matter has been lost entirely from the field.

Table 6.1 *Example: Field of Maize and Nine Indicators*

Indicator	Quantitative assessment	Interpretation
Rills within planted beds of maize (0.01m³ per 5m² of field)	Rill erosion of 26t/ha since the last field cultivation to prepare ground prior to planting two months ago	This rill erosion has occurred in the current season: probably most of it in very early season storms before the crop has germinated. Rills act to channel excess water and sediment – so the soil loss represented by the volume of the rill will only be a fraction of total soil loss from the field. (This observation is corroborated by the pedestals, suggesting an approximately 4:1 ratio between sheet soil loss and rills – about right for most fields.) Now, with weeds placed in the rills and the better cover from the maize, there will be little more additional rilling – maybe even some sedimentation
Pedestals in-field (2.5mm high)	Sheet erosion of 32.5t/ha in the last 3 weeks since weeding	This is a significant removal of soil during the middle growth period of the maize, indicating that the crop has given relatively poor cover to the soil. The erosion rate in the 3–4 weeks prior to weeding and after planting must have been just as high, if not higher, because of the poorer vegetation cover. The observer needs to enquire whether there were large rains then. If there were, then this suggests an annual sheet erosion rate of the order of 70–100t/ha
Tree mounds (150mm high)	Cumulative sheet erosion of 1950t/ha over the last 20 years	If distributed evenly over the 20 years, there would have been nearly 100t/ha/yr sheet erosion in this field. Erosion in the early years would likely have been less because the soil would have been in better condition. So this indicates a high long-term rate of erosion of 100t/ha/yr since deforestation, and a current rate of erosion of possibly 120–150t/ha/yr. These figures are slightly higher than those calculated from current sheet erosion (pedestal indicator) plus rill erosion – (70–100 + 26). The assumption of lower soil loss in early years may be incorrect – ask the farmer what was grown then and if the land had been kept bare or suffered major rainstorms
Boundary accumulation (10m³ per hectare)	13t/ha in the last year	This is a new grass path created just over a year ago. The grass has intercepted sediment and water from the field, and the accumulation has built up. But from these figures, it is apparent that about 90 per cent of the sediment has gone through the boundary, probably in the larger storms. Nevertheless, the boundary has succeeded in 'saving' 10 per cent of the loss, including some fine particles. Over time, the interceptive ability of this grassed path should get better, as the field slope reduces by the accumulation and the grass becomes more vigorous. Additionally, the deposited sediment will be fertile and so a better crop should grow – see next indicator
Sediment in within-field drainage ditch (0.001m³ per metre length of drain)	2.6t/ha since preparation of land two months ago	These sediments represent only the coarsest fraction of the soil that has moved across the upper part of the field slope. Field observation of its texture suggests that this fraction is only 10 per cent of the whole soil. Hence, this is evidence that a minimum of 26t/ha of soil was eroded to produce this material. It is a minimum because some of this same sand fraction may have remained in the field (and not caught in the drain), and some may have been washed out of the drain in very large storms. Because erosion selectively removes the

Table 6.1 *continued*

Indicator	Quantitative assessment	Interpretation
		fine particles, the actual amount of soil eroded in the two months must have been much larger than the 26t/ha calculation, which is not inconsistent with the 70–100t/ha from pedestals indicator
Soil depth (250mm deep at top of field; 500+ mm at bottom)	Sheet erosion loss of 250mm in 20 years; or about 160t/ha/yr	This reduction in soil depth, based on farmer estimates of original soil depth (but capable of corroboration by the field assessor on an adjacent site at same position on slope), occurred at the top of the field where maximum erosion has happened. However, some of this loss is 'cultivation erosion': ie the farmer has dug soil downslope. The field assessor needs to determine to what extent this direct intervention in land degradation by the farmer should be included. As there has been deposition at the base of the field (hence erosion is zero there), 160t/ha/yr would give an average sheet erosion over the field of 80t/ha/yr since the field was opened up
Within-field variation of crop growth	No quantitative assessment possible	Observations are consistent with soil having moved from the top part of the field to the bottom. This indicator is a measure of impact of land degradation, showing that crop growth on the 'eroded' soil at the top is significantly poorer than lower down the field where soil removal has been less, and very much poorer than on the lower boundary where there has been some deposition
Maize nutrient deficiencies	No quantitative assessment possible	The stunted, yellow plants at the top of the field are clear evidence of both poor growth because of lack of soil rooting volume and lack of sufficient nutrients and water. In that germination (as evidenced by plant density) was relatively uniform, the restricted growth only became evident once the plant had higher demands for nutrients and water. The purplish colours of the leaves in mid-slope is evidence of phosphorus deficiency. Phosphates are easily washed downslope by erosion; some may have accumulated in the deposited sediment (hence the good growth there) but most have been taken off in solution. The deep green of the plants at the lower end of the field indicates good water and nitrogen supply – much of this is accumulated from the higher parts of the field
Enrichment of sediment in a downstream hollow (5:1)	Five times as much clay in the hollow than in the soil from which it came	The hollow will have trapped a representative sample of water and sediment exiting from the whole field. As the puddle in the hollow dried out, the clays and other fine material (eg humus) settle out. The 5:1 enrichment indicates that the impact of land degradation processes is a very significant influence on the fertility of the field. Most of the sands are redistributed in the field, but the main fertile fractions are almost (except for a small amount trapped behind the grass path boundary) completely removed from the field. Future production will be affected far more than in proportion to total amounts of soil lost – the factor 5 suggests a crash in yields after only a few more years unless remedial measures are taken

Combining Indicators in the SRL Approach

Bringing all the information together in a common framework that puts the farmer-perspective to the forefront is encouraged throughout this handbook. Chapter 3 and Table 3.1 gave a model for field assessment in terms of the 'capital assets' of land users. These assets, divided into natural, physical, human, social and financial capital, provide a useful means of assembling all relevant items of information that have been identified. An abbreviated example based on a field of maize (Table 6.1) and a farmer (Plate 39) is provided in Table 6.2.

Using the SRL framework in this way thus enables a balanced view of the complexities of real farming. Nothing is simple. Apparently simple solutions such as added financial capital assets from the rich uncle may mean ambivalent outcomes – a ridger is good for preventing land degradation, but has further demands in needing oxen in a timely fashion. These demands potentially exacerbate land degradation when they cannot be met – in this case by convincing the neighbour to let him have ploughing done first. It is important that this sort of 'balance sheet' of how the farming situation changes the assets of a farmer to gain a livelihood is built in a systematic way. Later, in Chapter 8, a quantitative way (investment appraisal) will be used to bring this framework into an economic analysis. But for the moment, semi-quantitative and non-quantitative indicators provide a useful means of gaining a full impression of the land degradation situation.

Guidelines for Combining Indicators

Finally, in this chapter, some guidance is given on how to approach the challenge of combining indicators. It is a challenge because studies in land degradation have been bedevilled by reductionism. Approaches to measurement have usually been satisfied with single sets of observations rather than the approach advocated here. Yet, the example in Table 6.1 demonstrates that a comprehensive view of the effect of the history of land use can be gained if the pieces of information are set side by side. Furthermore, the example in Table 6.2 shows how combining indicators can give deep insight into current effects of the land user on land degradation. The field assessor must, therefore, have an open mind, observant eyes, and the qualities of a detective.

It is difficult to provide specific guidance for all situations – there are many permutations of possible land degradation and land use conditions, and hence many possible interpretations. Therefore, in the following two sub-sections, suggestions are made for (a) the approach to adopt in the field, and (b) how to put the indicators together in a semi-quantitative form for initial inspection.

A checklist for the field

It is important to make a careful reconnaissance of the field site to note all the pieces of evidence of both land degradation processes and their impact. The

Table 6.2 *Combining Indicators in the SRL Framework for a Field of Maize: How Land Degradation is Affected*

Capital asset	Positive effects of change in capital	Negative effects of change in capital
Natural	Farmers have planted field boundaries, against which some soil accumulates (13t/ha in the last year). As these boundaries are enriched by planting of fruit trees and other economic species, their effectiveness in accumulating natural capital will increase	Deforestation has led to a substantial loss of natural capital. The soil is now eroded; its water-retaining properties are deficient; and the overall stocks of biomass and plant diversity are much reduced. Biophysical indicators (Table 6.1) summarize the effects
Physical	The farmer (Plate 39) has only a hoe for cultivation. This means he cannot extend his cultivation to larger areas. Instead, then, he has to intensify land use on small plots and use the benefits of multiple cropping to limit the need for more tools and equipment. This is good for conservation, but hard for the farmer	Poverty means little opportunity to accumulate further means to manage the land resources. In effect the farmer is confined to cultivating simply with no means of physical conservation. In addition, distance from markets and physical infrastructure gives little opportunity to grow high value crops for sale
Human	The farmer has a wealth of indigenous knowledge, handed down from his father. This includes the small drainage ditch across the field to protect from runoff. He tells us about techniques he knows of composting and of building small terraces. These would be excellent to control land degradation. But in the pressure to grow the maximum amount of maize for home consumption, much of this knowledge is not applied	Old age and ill-health in the family (his wife is very sick) means that farming practices must be minimized if enough land is to be cultivated for sufficient food to be grown. Human capital limitations determine that the farmer's time horizon is short, and that there can be little investment in the future – except those activities which demand least labour (planting field boundary) and those that are essential for survival (cultivating maize)
Social	Family and clan ties have enabled the farmer to call on relatives and clansmen to get the field ploughed early in the rainy season. This has enabled timely planting and minimized the risk of erosion because of poor vegetation cover. The maize crop is looking good (Plate 39) mainly because of this communal effort in planting the seed on time	Family and clan ties also mean that part of the crop has to be given over to other members of the social network. To do this, the farmer has to take off-farm employment to supplement income. He cannot then devote time to carrying out protective measures such as managing the runoff safely, and to dealing with the maize nutrient deficiencies which manifest themselves in late season
Financial	A rich uncle in the capital city remits enough money for the farmer to buy an ox-drawn ridger. Next year he can plough across the slope, with planting undertaken on conservation ridges that prevent further land degradation … but….	No bullocks to pull the ridger. After negotiating with a neighbour, he gets enough cash to hire the animals for ploughing next season. However, it is now late in the new season because the neighbour wanted, understandably, to plough first. The animals are exhausted, and the crop planting is a failure. More land degradation

Note: (See Sustainable Rural Livelihood Model in Table 3.1 and field data example in Table 6.1 with added information from farmer interviews.)

following checklist is for general guidance only. Like any good detective, the field assessor must follow up any interesting leads, especially those initiated by comments from the farmer.

1   Map out the field slope as a sketch, noting the position of any obvious features such as gullies, rills, tree mounds, boundary walls. Any differential crop growth or obvious nutrient deficiencies should also be noted on the map.
2   Obtain the history of land use: when the plot of land started to be used, crops grown, any change in land use, subdivisions of the land, and similar important events that could have a bearing on land degradation. (These events should later be set alongside the field measurements to ascertain whether they correspond with observations.)
3   Determine any significant events: landslides, exceptionally heavy storms and soil wash, dates when trees were cut down.
4   Note any particular farming techniques that may have implications for land degradation (eg ridging practices across/down the slope; hand cultivation downslope).
5   Then, with the map in 1 above, and preferably accompanied by the farmer, go through the indicators of the processes of land degradation:
    • soil losses from single places (eg tree mounds, pedestals, soil depth);
    • soil losses from small parts of the field (eg rills, armour layer);
    • soil losses from large parts or the whole of the field (eg gullies; differences in soil depth between degraded field and non-degraded; or averages over the field of previous items such as tree mounds);

    • sediment accumulations and their enrichment/texture within the field (eg drainage ditches, against an in-field tree);
    • sediment accumulations and their enrichment/texture at the base of the field (eg boundary accumulations);
    • sediment accumulations and their enrichment/texture outside the field (eg clay enrichment in hollows).
6   Then, with the farmer (most important this time), determine the indicators of the impact of land degradation:
    • observation of current plant growth (eg within-field differences);
    • actual measurements of different sizes of plants;
    • list known nutrient deficiencies observed;
    • estimate, with farmer, likely yields from different parts of the field;
    • obtain historical yields, and observations on how plant growth has changed.
7   Compile a comprehensive table of indicators and results, looking for trends, consistency and areas where there is broad agreement in the scale of degradation.
8   Return to the farmer with your account of the comprehensive picture, and get his/her evaluation of your diagnosis.

A semi-quantitative assessment

Assessment so far by combining indicators has attempted to use absolute (scale) levels of land degradation, such as tonnes of soil per hectare. With the approximate nature of the techniques of assessment, this can be misleading unless careful precautions ('health warnings') are taken. To say that exactly 126t/ha/yr of soil loss has occurred is folly, implying that it was more than 125 and less than 127. This degree of exactitude is unjustified. If it is suspected that someone may take these absolute figures (as has often happened) to use them as precise evidence of the level of degradation, then it may well be better not to give the figures in the first place. The alternative is a semi-quantitative assessment.

'Erosion Hazard Ratings – EHRs' (see Annotated Bibliography – Appendix IV) are one example. The factors of erosion – slope, soil type, vegetation cover and rainfall – are rated on a numeric scale, usually one to five in severity of likelihood to cause erosion. Then these individual factor ratings are combined, either through a scoring system or through a simple model, to give an overall hazard rating. This is not an actual measure of land degradation, but a prediction of potential land degradation according to the environmental factors that encourage it. Such assessments have been widely used for broad-scale planning purposes. They are simple to develop and easy to visualize since the results are usually presented in the form of a map (eg Figure 6.2). EHRs are not, however,

particularly useful at the detailed field level, or for developing a farmer-perspective approach.

Instead, Malcolm Douglas in his *Guidelines for the Monitoring and Evaluation of Better Land Husbandry* (see Appendix IV) has suggested simple scoring techniques for seriousness of indicators of land degradation (and conservation effectiveness). The reader is referred to this 28-page publication for more details. However, it is perfectly appropriate to develop one's own scoring system. Provided that it is consistently used, it can be a good way of combining indicators to get a more comprehensive view of land degradation.

Tables 6.3 and 6.4 give two of the more commonly used examples that combine observations of a number of separate indicators.

Douglas also suggests a three-point scale for the effectiveness of conservation for (a) crop management, and (b) soil management. The 'effectiveness of conservation' is essentially a composite view of both direct and indirect field interventions by the land user. They include how effec-

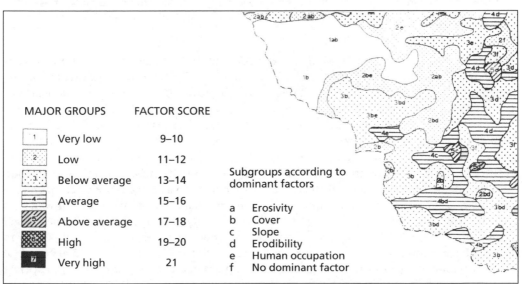

Figure 6.2 *Extract from Erosion Hazard Assessment for Zimbabwe*

Table 6.3 *Sheet Erosion*

Ranking	Degree	Description
X	Not apparent	No obvious signs of sheet erosion, but evidence of minor sheet erosion may have been masked, for instance by tillage
0	No sheet erosion	No visual indicators of sheet erosion
1	Slight	Some visual evidence of the movement of topsoil particles downslope through surface wash; no evidence of pedestal development; only a few superficial roots exposed
2	Moderate	Clear signs of transportation and deposition of topsoil particles downslope through surface wash; some pedestalling but individual pedestals no more than 50mm high; some tree and crop roots exposed within the topsoil; evidence of topsoil removal but no subsoil horizons exposed
3	Severe	Clear evidence of the wholesale transportation and deposition of topsoil particles downslope through surface wash; individual pedestals over 50mm high; extensive exposure of tree and crop roots; subsoil horizons exposed at or close to the soil surface

Table 6.4 *Rill Erosion*

Ranking	Degree	Description
0	No rill erosion	No rills present within the field
1	Slight	A few shallow (< 100mm depth) rills affecting no more than 5 per cent of the surface area
2	Moderate	Presence of shallow to moderately deep rills (< 200mm depth) and/or rills affecting up to 25 per cent of the surface area
3	Severe	Presence of deep rills (up to 300mm depth) and/or rills affecting more than 25 per cent of the surface area

tively crops protect the soil as well as the use of fertilizer and specific 'land husbandry' practices. Douglas's tables (adapted for this handbook) give a very useful checklist of land user practices, as well as bringing together a diverse number of farmer activities into a comprehensive picture of the land degradation potential.

To gain a composite view of the influence of crop management on land degradation, the six crop management indicators (Table 6.5) are scored one to three. The minimum score is six, indicating that crop management practices minimize the risk of land degradation; the maximum is 18, indicating extreme danger of rapid degradation. To bring the composite view back to a 1–3 scoring scale, divide the sum of the scores by the number of indicators – in this case divide by six. The 'conservation effectiveness' can then be interpreted comparing different crop management regimes in their likelihood to contribute to land degradation. So, a total score of eight that gives an average score of 1.3 would be interpreted as 'crop management practices are largely effective in limiting the danger of land degradation and could help to rehabilitate existing degraded land if implemented'. Locally appropriate descriptions should be developed for ranges of scores, for example:

Table 6.5 *Crop Management Considerations*

Crop management indicators	Conservation effective Score 1	Conservation neutral Score 2	Conservation negative Score 3
Change in percentage ground cover by the growing crop	At least 40 per cent cover of soil achieved by crop within 30 days of the start of the rainy season	Little increase in ground cover provided by crop between fallow and growing crop	Decrease in ground cover – remains below 40 per cent for most of the growing season
Intercropping/ relay cropping	Cropping practices lead to improved ground cover and/or increase in the ratio of legumes (N-fixing) to non-legumes	No change in intercropping or relay cropping practices	Cropping practices lead to reduction in ground cover and/or decrease in the ratio of legumes (N-fixing) to non-legumes
Spacing/ planting density	Ground cover improved through closer crop spacing and/or increased plant density	No change in plant spacing and density	Ground cover reduced through wider crop spacing and/or decreased plant density
Improved seed/ planting material	Adoption of improved seed/planting material results in improved biomass production and better ground cover	No change in crop biomass and ground cover	Adoption of improved seed/planting material results in decreased biomass production and inferior ground cover
Fertilizer and/or organic manures	Increase in fertilizer and/or organic manures results in more biomass production and better ground cover	No change in quantity of fertilizer and/or organic manures used for crop production	Decrease in fertilizer and/or organic manures results in less biomass production and poorer ground cover
Crop residues	Crop residues incorporated into the soil or retained on surface as protective mulch	Not applicable	Crop residues burnt or fed to livestock

1.00–1.49   No land degradation danger; good rehabilitation potential

1.50–1.99   Land degradation slight; good possibility of rehabilitation

2.00–2.49   Moderate danger of land degradation – particular practices have specific problems

2.50–3.00   Land degradation hazard high to very high – all practices contribute to danger

For soil management considerations in Table 6.6, a minimum score of three and a maximum of nine is possible against the three indicator variables. The same procedures apply to interpret these scores in terms of overall contribution to land degradation status.

Such tables should be adapted for the specific circumstances of each field and the different types of land use. Once developed for a local area, they can provide excellent ratings to determine the specific danger of different types of land use. Furthermore, they can be used to assess proposed interventions alongside existing practices to see if land degradation status will be unduly changed. Such semi-quantitative techniques, therefore, provide both a current view and a predictive means to monitor land degradation status.

Table 6.6 *Soil Management Considerations*

Soil management indicators	Conservation effective Score 1	Conservation neutral Score 2	Conservation negative Score 3
Soil organic matter	Interventions enhance soil organic matter through, for example: • incorporation of crop residues • application of at least 3t/ha/yr compost and/or animal manure • application of at least 5t/ha/yr of fresh green manure (eg *Leucaena*)	Interventions only maintain soil organic matter levels by: • grazing livestock on crop residues *in situ* • application of compost and/or animal manure at rate below 3t/ha/yr • application of fresh green manure at rate below 5t/ha/yr	Interventions fail to maintain soil organic matter levels: • removal or burning of all crop residues • no application of compost and/or animal manures • no application of green manure – all biomass removed as fuel or fodder
Soil chemical properties	Interventions replace lost soil nutrients through: • application of compost and/or animal manure • use of N-fixing species in crop rotations and intercropping, or in N-rich green manures and hedgerows • enriched fallows • chemical fertilizer (as a supplement, not a substitute for organic manures)	Traditional low input fertility management practices capable of achieving low levels of nutrient replenishment, through: • short bush fallow • tethered grazing of livestock within farm plots on crop residues and weeds • retention of a few scattered trees on the croplands	Poor practices that continue the depletion of soil nutrients through: • continuous cultivation of cereal and root crops • burning of crop residues • little, if any, use of compost, organic manures or chemical fertilizer
Soil physical properties	Interventions maintain and enhance topsoil structure through: • minimum tillage • planted pasture and enriched fallows • incorporation of crop residues, compost, animal manure, green manures and tree litter	Traditional low input practices neither combat nor promote physical degradation of the soil, through: • partial tillage • short bush fallow • retention of a few scattered trees on the croplands	Poor practices continue physical degradation of the soil, through: • excessive tillage • continuous cultivation • no incorporation of organic matter • trampling by people and livestock

Consequences of Land Degradation for Land Users

A Game of Winners and Losers

Often land degradation is perceived as having only negative effects for land users and society alike. These effects are far-reaching with off-site costs (referred to in Chapter 2) extending well beyond the site where land degradation has occurred. The effects of land degradation may be experienced by future generations, affecting how they can use the land. Box 7.1 lists some actors typically affected by land degradation.

However, this does not disclose the full picture. A cynic might say that 'land degradation is a game of winners and losers'. There is an important grain of truth here: some people do gain from land degradation, while others lose. It is a fact, unpalatable to many, that land degradation can bring decided advantage to some people some of the time. If land degrada-

tion were only negative all the time, and if it were to mean that consistently everyone loses cash and resources, then undoubtedly society would have derived successful ways to combat it. Thus, it is important that the field assessor is alert to the potential gains that may accrue to some parties as a result of land degradation, as this is critical to an understanding of the attitudes of land users to land degradation. The field assessor should try to gain a clear picture, preferably in quantitative terms, of how the various actors may derive costs or benefits from the continuation of land degradation, and costs or benefits from the control of land degradation. Such an understanding is key to the design of sustainable programmes of rehabilitation, which will be the subject of the final chapter (8) of this handbook.

BOX 7.1 WHO IS AFFECTED BY LAND DEGRADATION?

The user of degraded land, through loss of current production and more difficult farming operations

The person who uses the piece of land immediately downhill from the degraded site may face landslides or floods, even though their land may not, itself, be degraded directly by its user

The organization that supervises the hydro-electric dam that becomes sedimented

The port authority that has to pay for dredging channels for ships and the water authority that has to install expensive purification works

The national economy may lose through collecting fewer tax revenues from land users, and by a decrease in the value of the stock of natural resources

---

So, who has a stake in land degradation continuing, or at least would suffer some disadvantage if more effort were put into land degradation control? Although it is impossible to generalize, and each case presents unique combinations of actors and biophysical processes, the 'winners', those who gain from land degradation, can be drawn from the same categories of people as those who lose as a result of land degradation. Examples of possible winners include:

- **Immediate land users** at the site of the degradation: if land users were to have to spend more money to control the land degradation than they would gain from the increase in production consequent on conservation, then they would, on balance, be winners in allowing land degradation to continue.
- **Neighbouring land users:** if land degradation were to cause scarcity of productive land, then neighbours with land that remains usable may find that land prices increase or rental opportunities become available.
- **Downstream land users:** if they gain rich eroded sediments from uphill, then the productive value of the land would similarly gain.
- **Authorities** responsible for dealing with the impacts of land degradation: in a perverse way, land degradation can increase their power and importance. If, say, electricity supply were interrupted because of damaged hydro-electric installations or sedimented waterways, the authorities may claim powers to evict upstream land users and alienate land for forestry over which the authority

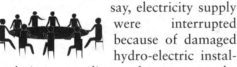

may gain production benefits.

- **Governments:** many, if not most, developing country governments are maintained in power by the poor rural peasantry. Their chief opposition tends to come from urban elites. Keeping the peasantry poor (on degraded land) can mean political advantage and continuation in power. Diverting resources from agriculture to keeping the urban masses contented is another way of not controlling land degradation, which may help to keep the more volatile political ambitions of urban people in check.
- **Scientists and professionals:** land degradation is a subject that concerns researchers, teachers and practitioners in every country. It directly or indirectly employs many thousands of people. If it were not for land degradation, this book would not have been written.

Land degradation is by definition (see Chapter 2) the **aggregate** diminution of the productive potential of the land. Society as a whole loses, but within that aggregate loss, some people gain. The paddy fields of Sri Lanka are enriched by sediments from the fields of farmers above (see Plate 40). It is essential to recognize this in order to appreciate the appropriate points of intervention in land degradation control and to consider those persons who would lose if land degradation were to be controlled.[1] However, the focus of this handbook is on a farmer-perspective, and so it is the analysis of field indicators within a framework of the consequences of land degradation for the immediate land users of the degraded land that is discussed in the following sections.

Outcomes of Land Degradation

One of the best-researched areas of land degradation outcomes for land users is in the impact of soil erosion on yields of crops – called erosion-induced loss in soil productivity. By combining indicators of both the *process* of degradation and its *impact*, the field assessor can develop scenario outcomes. Before describing how to develop such scenarios, an understanding needs to be gained of the principal variables involved.

For erosion-induced loss in soil productivity, there are three major variables:

1   **Soil erosion** – represented as a cumulative amount of soil lost; that is, as soil loss over a unit of land area accumulated over specified periods of time. This can be expressed as a **rate** of soil loss (eg t/ha/yr) or as a **cumulative amount** for a longer time period such as 20 years (eg t/ha over 20 years).
2   **Time** – the length of time over which soil loss and declining yields are measured, and the time over which scenarios are constructed (usually 20 years or more).
3   **Crop yield** – represented as yield of a major crop, such as maize over one growing season (eg kg/ha).

There are three pair-wise permutations of these three variables: erosion–time; erosion–yield; yield–time. Each permutation gives us interesting insights into the outcome of land degradation and the effects this has on land users (see 'Constructing Scenarios: Theoretical Perspectives' below). The terms **sensitivity** and **resilience** introduced in Chapter 2 enable description to be made of the status of the land and its potential for dynamic change and effect on land users. Consider a number of combinations of **erosion rate** (erosion–time) and **impact** (yield), and compare these with Table 2.2, the sensitivity–resilience matrix for land degradation:[2]

1   There is little soil erosion, and what little there is has almost no effect on yields – low sensitivity and probably high resilience, if restoration of yields is easy
2   There is little soil erosion, but what little there is has a large impact on yields – low sensitivity and low resilience, if yield restoration is difficult
3   There is much soil erosion, but there is little impact on yields – high sensitivity and high resilience
4   There is much soil erosion, and a large impact on yields – high sensitivity and low resilience

Obviously there is a continuum between these four states, with the majority of cases occurring somewhere between high/low sensitivity and resilience. They do mean for the land user that very different land use and conservation practices present themselves as being rational.

These observations are only snapshots of the sort of outcomes that could be predicted in determining the consequences of land degradation for the land user. They bring in additional variables, which the field assessor can determine, such as:

•   Inherent soil fertility: it makes a lot of difference to many outcomes whether a soil starts in good condition (then it may well be worth expending effort to keep it good) or in poor condition (where effort may have no discernible return in improving the soil or giving better yields).

- Resource endowments of the land user: this is a complicated subject, but to demonstrate, simply consider two farmers, one with a little land, the other with a lot. The land-poor farmer will need to retain the quality of the land, if this is the only source of income, usually by intensifying production. The land-rich farmer can switch production to other fields, leaving the eroded land to regain fertility through fallowing. This is a case where poor farmers perhaps best serve conservation. The very opposite may pertain for resource endowments in capital. The rich farmer can afford to implement physical conservation measures, by hiring labour and equipment; the poor cannot.

Constructing Scenarios: Theoretical Perspectives

The three outcome permutations of erosion-induced loss in soil productivity are considered here in turn. Each of the following three sub-sections illustrates theoretically how scenario predictions can be derived from typical data sets that could be gained by field assessment. (The results that illustrate them in the graphs come from work completed for the UN Food and Agriculture Organization on food security in Africa; see Appendix IV: Annotated Bibliography.) The practical ways in which these relationships could be developed are considered later in this chapter, in 'Constructing Scenarios: Practical Issues'.

Erosion–time

A typical set of erosion–time relationships for different land uses is given in Figure 7.1, derived from Zimbabwean data. At a broad level of analysis, it can be seen clearly how the extreme curves (for bare soil and early-planted maize) progressively diverge over time. For the field analyst this is useful, because it shows over time how much soil is 'saved' by the maize that gives the best cover. Indeed, the *rate* of divergence progressively increases as the bare soil gets more and more degraded. This is a common phenomenon on most soils, that degraded conditions lead to even

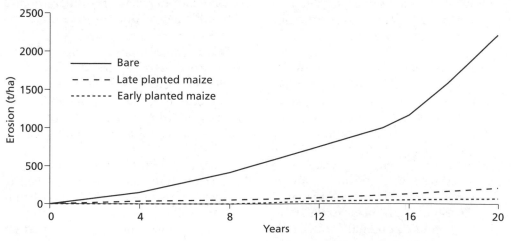

Figure 7.1 *Cumulative Erosion under Different Land Uses*

more degradation, while land use that has biological forms of conservation tends to become more resilient as organic matter levels increase. It is these sorts of changes and the dynamic interactions with land use that are so critical to both the farmer and the field assessor of land degradation.

Erosion–yield

Figure 7.2 shows the way in which yields decline with cumulative soil loss – the sort of soil loss that could accumulate over a number of years depending on the rate of land degradation. The least resilient soil, the Phaeozem, has the sharpest decline in yields. This means that per unit of soil loss, the greatest impact of erosion is with this relatively productive soil. On the other hand, a Phaeozem is not especially sensitive – it does not erode easily. It would be worthwhile for the land user to concentrate efforts on preventing erosion with this soil, because in this way maximum yields can be 'saved' and a good soil retained in reasonable condition. The opposite is true of Ferralsols – they do erode easily (that is, they are sensitive), but per unit of soil loss, this has modest impact (they are resilient). For the land user, this means that it may not be worth worrying

too much about land degradation, because only when erosion becomes very severe is there a substantial impact on yields.

Yield–time

Figure 7.3 demonstrates for a Luvisol how different standards of management, as assessed by degree of plant cover, affect yield–time scenarios. The Luvisol is one of the most strongly differentiated soils in this respect – good management as represented by good cover can keep yields continuously high, while poor management as represented by bare soil lets the yields crash. Again, the implications for land users may be drawn. Indeed, the aspect of time is very useful because it relates directly to the concerns of a farmer.

The yield–time scenario may be interrogated to ask 'how long can I use this soil (or piece of land) before yields decline to a level where it is not worth my continuing?' Table 7.1 gives soil life spans for some South American sites. The implications of the years of useful life of the soil are potentially huge for the land user. For the Ferralsol and Acrisol, it makes little difference – however the soil is managed, degradation will mean that it will never be usable economically for more than four

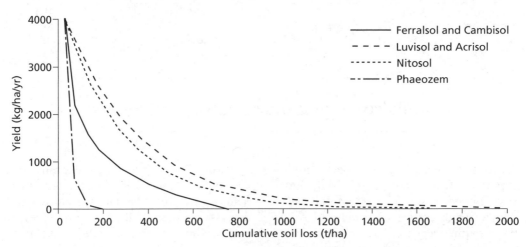

Figure 7.2 *Erosion–Productivity Relationships for Different Soil Types*

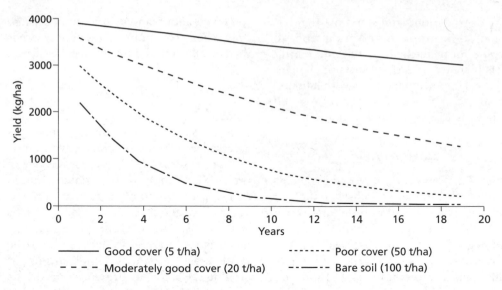

Figure 7.3 *Maize Yield Decline with Erosion for Luvisols*

years continuously. On the other hand, a Cambisol and Phaeozem can be used even to moderate levels of management for a generation or more. The Nitosol by contrast is absolutely and robustly sustain-

able provided that management is consistently good – otherwise, after only a few years the yields decline to such an extent it is not worth using any more.

Table 7.1 *Years Taken for Different Soils to Reach a Critical Yield Level of 1000 kg/ha/yr with Continued Erosion*

Management level	Ferralsol	Acrisol	Luvisol	Phaeozem	Cambisol	Nitosol
Good cover	3	4	93	200	210	950
Moderate cover	2	3	23	65	42	19
Poor cover	1	2	9	7	23	4
Bare soil	1	1	5	3	9	3

Constructing Scenarios: Practical Issues

As noted already, these are complex scenarios. The outcomes of land degradation involve many factors, only some of which are directly measurable, but even these require repeated measurements over many years in order to construct reliable

relationships with land degradation. Consequently, scenarios constructed by the field assessor from field measurements and discussions with land users will represent best estimates. Nevertheless, isolating erosion–yield–time relationships is an excellent way of bringing together the outcomes of greatest importance from a farmer-perspective. So, it is worth the effort, even if estimates or guesses must be made to fill in missing items of data. These estimates or guesses must, of course, be given prominent 'health warnings' (see Chapter 1 of this handbook).

STEP 1: Targeting time-series sources of information

The objective of time-series data (Table 7.2) should be to develop a database of qualitative and quantitative changes in land degradation and associated factors. These factors should be both causative (eg major floods) and impact-related (eg triggering migration).

STEP 2: Plotting yield–impact data over time

The objective here is to develop a curve of yield over time. Both the starting yield and the current yield are important. The first is a reflection of what the yield would have been on first opening up the land. It is best to try to isolate the effect of fertilizers by considering only yields at low input levels.

Sources of information on yield impact include the following:

- From time-series data above, trends in yields should be plotted. Land degradation is, however, only one element in changing production. Fertilizers or new crop varieties can mask the effect of land degradation.

- *In situ* yield differences, using soil variables as a proxy for time. An example of this is to measure the differences in yield between a degraded and a non-degraded field; or a part of the field where erosion has reduced the depth of topsoil and another part where accumulation has occurred. In both cases, soil depth or other soil variable acts as a surrogate measure of land degradation over time.

- Using soil variables such as depth, develop a spatial database of current crop yields over the fields being surveyed and compare these with actual yields reported by the farmer or directly measured.

Table 7.2 *Time-series Data*

Source	Data Type
Farmer	Records of yields, knowledge as to how things have changed in the field and the farmer's response to these changes. It is good to develop a time-line of land use and practice for a field, starting from when it was originally opened up for land use
In-field	Observations of erosion rate indicators for different time periods (see Chapter 4). Longer-term indicators such as build-up of sediment behind boundaries, the past position of gully heads or the development of tree mounds are especially useful. Farmers' knowledge is again valuable
Community	Observations of major changes and events, such as droughts, famines, migrations (and other population movements). While not directly quantifiable, such evidence supplements and may verify field evidence of change
Official records	Delivery of crops to market, purchase of fertilizers, field inspections

STEP 3: Bringing in supplementary information

The most common and probably the most useful supplementary information is 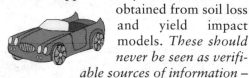 obtained from soil loss and yield impact models. *These should never be seen as verifiable sources of information* – model outputs are only as accurate as the quality of the data that went into them. Taking data from one place to very different conditions of soil, climate, slope and land use has been shown to lead to enormous errors. Nevertheless, with care, models can give useful supporting data, especially if the outputs of the models are calibrated and verified with actual field data. It is outside the scope of this handbook to detail the available models and to show their application. Other manuals exist that do this. Models used to predict soil loss and yield impact include:

- USLE – Universal Soil Loss Equation: a wholly parametric model that has been used by the US Natural Resources Conservation Service for many years. It works poorly as an *absolute* predictor of soil loss for tropical conditions. However, model predictions of soil loss are usually in the correct rank order, and so it can be used to show what land uses are more hazardous to land degradation than others.
- SLEMSA – Soil Loss Estimation Model for Southern Africa, developed in Zimbabwe as a lower cost alternative to USLE, requiring fewer data. Because it is derived from sub-tropical conditions, SLEMSA seems to have better predictive capability than the USLE for steep lands and tropical soils. But again, care needs to be exercised in using the model outputs as absolute predictions.

- PI – Productivity Index. This is a fairly simple model based on a very limited range of soil variables such as soil depth, which plots yield changes on the assumption that a single variable will usually be the single most limiting factor.
- EPIC – Erosion-Productivity Impact Calculator. This US model is essentially a USLE extension into productivity, involving a large number of sub-models. Unless EPIC has been developed and verified for the particular area being surveyed for land degradation, the amount of effort required to make it work is unlikely to be rewarded with usable results.

STEP 4: Constructing erosion–yield–time curves

The three pair-wise permutations of  erosion, yield and time each have their main functions and uses. It is suggested that single data point plots are plotted on three graphs. Approximate lines can then be interpolated on the following basis:

- Land degradation tends to plot to an exponential function. In other words, as land becomes more degraded, its rate of degradation increases (see Figure 7.1).
- The difference between land uses can be assigned according to a cover curve (Figure 7.4), which expresses the proportional change in erosion consequent on a change in mean vegetation cover (see Appendix IV, Lal, 1994). So, if a land use has a mean vegetation cover that gives an erosion percentage of 1 per cent (ie 1 per cent of the

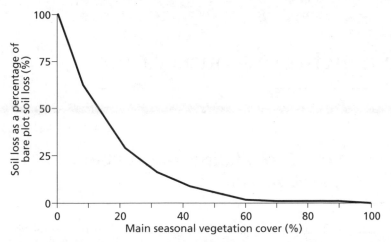

Figure 7.4 *Generalized Erosion–Cover Relationship*

does obviously need to concentrate on the losers. The loser is more likely to be the marginal person in society, as well as the land user living on the most difficult parts of the landscape such as steep slopes and erodible soils. However, as this chapter has highlighted, the field assessor needs to keep an analytical eye on the winners – the people who have everything to gain from degradation continuing. If, unlike the losers, the winners are the more privileged in society, they are more able to maintain the status quo and ensure that land degradation control measures are ineffective.

erosion that would have occurred if the land had been bare), while another land use has a percentage of 30 per cent, then the second land use has 30 times the erosion of the first.

In the final analysis, these scenarios are merely forums for bringing together diverse sources of data into a form that has meaning both to the land user and the professional. The ultimate objective is always to gain a realistic assessment of the consequences of land degradation to land users. If land degradation is truly a game of winners and losers, the field assessor

In the final chapter of this handbook, the information so far gained is used to examine and analyse the potential benefits of the reverse side of the 'land degradation coin' – conservation and rehabilitation.

The Benefits of Conservation

Extending Land Degradation Assessment into Conservation

The main practical and field-level purpose of land degradation assessment is to determine what measures of conservation are the most appropriate, and best meet the differing objectives of all parties affected by the degradation. Conservation is the next logical move from considering the consequences of land degradation for land users. Appreciating how all this information might be turned to productive benefit for the land user is an exciting challenge for the field assessor.

The identification and measurement of land degradation are essential steps in developing conservation strategies. While these measurements may provide useful information about environmental change, unless applied to reduce land degradation they serve little practical purpose. So, land degradation assessment should not be seen as an end in itself – it is a means to achieve a practical and useful outcome for a specified user of information. Users include planners, other professionals, development practitioners, field staff, farmers or the rural poor. Users may come from government agencies, international organizations, local and international non-governmental organizations, companies, research institutes and individuals. Their objectives and purposes for wanting land degradation information may be very different. Thus, it is vital that any programme of land degra-

dation assessment has a clear idea of who the client is, what the client wants and how the information is to be used. In this handbook the particular emphasis is on farmers again, and how they may derive benefits from applying practices of land degradation control.

Meeting multiple objectives is a desirable aim. The classic *win–win* scenario is where the control of land degradation not only achieves a benefit to society, but also brings immediate support to the land user. One of the problems of conservation in the past is that these dual objectives of providing for both society and the individual land user were normally seen as incompatible. Typically, conservation was enforced on farmers, who then had to carry out additional and costly works to implement the recommended measures. So, conservation was always viewed as being a cost to the land user in extra labour, additional effort and more trouble. Farmers would justifiably say, 'Why should I live a poorer life, so that people downstream and in the towns can live a richer life?' It is not that land users are anti-social, but that in many cases their implementation of land degradation control measures on behalf of society represents a major cost to them and their families. This cannot be a sustainable position – nobody works in order to become poorer. But if, instead, land degradation were reduced *and* the needs of the farmer were met, this would be a sustainable solution – sustainable biophysically and economically, because everyone benefits. This is neat in theory, but difficult to put into practice. Good field-level land degradation assessment is a necessary condition to identify such sustainable solutions. When combined with an assessment of the socio-economic conditions of the land user, the field assessor is in a position to evaluate different conservation strategies.

Potential users of the information from land degradation assessments, such as professionals and development practitioners, will want to know whether the land user addressing land degradation will actually gain a benefit. This is a crucial question that underwrites much of what we have reviewed in this handbook. Land degradation assessment, including knowledge of its impact on the land user, is the primary entry point to estimating the **cost** of land degradation. By preventing this land degradation through measures of conservation, a **benefit** is derived for the land user in terms of yields and easier farming practices. The benefit is, in effect, the amount 'saved' above a baseline of continuing degradation, the 'without conservation' line in Figure 8.1. So even if yields simply remain constant, there is a benefit as represented by the shaded part of Figure 8.1.

In simple terms, if the value of the 'saved' yield exceeds the costs of implementing the conservation, then land degradation control through these means is potentially worthwhile to the land user. The technologies have a good chance of adoption.

The key to making such predictions of likely acceptance of conservation is being able to value accurately from a **farmer-perspective** the following:

- Costs of the conservation measure: this must include not only direct costs, such as the materials for a technology, but also indirect costs such as the amount of land taken up by the measure and the activities which cannot now be undertaken because of the time it takes to do the conservation.

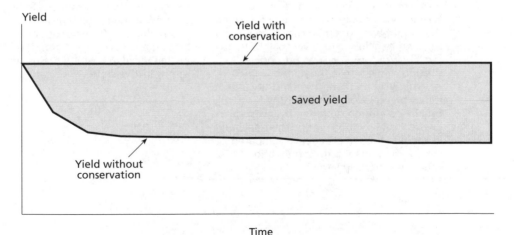

Figure 8.1 *The Benefit of Conservation*

- Benefits of the conservation measure: this must include both the direct benefits in getting increased crop yields (or maintaining existing yields, when otherwise they would have fallen without conservation), and the indirect benefits such as additional products, for example wooden poles from contour hedgerows.

The field assessor must try to assemble as much of this information as possible in order to move land degradation assessment on into a practical tool for determining what conservation measures are going to have a good chance of success.

Success has to be judged not only biophysically in terms of amounts of 'saved' soil, but more importantly economically, in terms of the net benefit (or cost) to the land user.

Sensitivity and resilience are measures already introduced in Chapter 2. They are used to determine the vulnerability of a landscape to degradation. They can equally be used to assess the likely effectiveness of conservation strategies. In Chapter 7 the impact on yields of each cell in the sensitivity–resilience matrix was assessed. The likely approach to conservation by the land user in each situation is summarized in the matrix in Table 8.1.

Table 8.1 *How Sensitivity and Resilience Affect Conservation Decisions*

	High sensitivity	*Low sensitivity*
High resilience	**Much soil erosion; little effect on yields** Conservation would have very doubtful payback to the farmer. On-site impact is low, but off-site impact might be far greater. It could be a case where society may want to provide subsidies to the land user to protect the land	**Little soil erosion; little effect on yields** Simple, low cost conservation that relies on biological means might be worthwhile to utilize the properties of high resilience
Low resilience	**Much soil erosion; large effect on yields** Two outcomes may occur. First, although soil erosion is high, there might be little to be gained by the land user in addressing the problem, especially if the inherent soil fertility is poor. A typical strategy for a land user might be to use it quickly, then leave the land to fallow until, eventually, natural fertility is restored. This occurs under shifting cultivation in the humid tropics. Alternatively, if inherent soil fertility is good, it may be very worthwhile for the farmer to address erosion problems by investing in physical conservation works. Their payback in keeping the soil productive compared with a situation with no conservation could be extremely large	**Little soil erosion; large effect on yields** Since yields are difficult to restore, there is every reason for the little soil erosion that might occur to be controlled. Some combined physical and biological strategy, such as ridging and cover crops, could well be beneficial in keeping the soil in its productive state

Typical Benefits of Conservation

At the outset, it is important to have a grasp of all the benefits that conservation may bring to the land user. These range from the immediate and direct right through to the remote. It is also vital to appreciate that farmers do not often implement conservation all in one go (see Box 8.1). Any assessment should attempt to capture the full range of benefits and ways of implementing conservation, otherwise some items that have great significance for farmers may be missed. A simple typology and two examples of each type are shown in Table 8.2, which can be used to develop a checklist for the field assessor when discussing benefits with farmers and carrying out field observations.

Similar listings of costs should also be constructed. As seen in the example in Box 8.2, the costs incurred and the opportunities forgone by implementing a conservation measure may be offset by the benefits, both in number and in value. For example, *Gliricidia* hedges take up 9 per cent of the area of the field but they also provide extra income opportunities from the sale of poles and firewood. In such complex cases, an understanding of costs and benefits should immediately alert the assessor to potential problems (such as poor rates of adoption, failure to maintain conservation measure) or to potential advantages if this technology were to be promoted.

Many, if not all, of these benefits and costs can be valued in financial terms, thereby laying the groundwork for cost-benefit analysis. However, care must be taken to avoid any element of 'double-counting'. For example, it may be unjustifiable in economic analysis to accept both an increased value of land as a long-term benefit and increased future yields. The increase in the value of the land may incorporate the better yields. Thus, once listed, costs and benefits should be carefully examined, arranged and codified for particular analytical purposes.

Box 8.1 Incremental Conservation

Many conservation techniques can be implemented incrementally – that is, a little at a time. Land users will implement those elements of a conservation strategy that make the best use of the time, labour and money that they have available. For example, in one year a land user may construct just two barriers across a slope, separated by a greater than optimum distance. In the following year a third barrier may be constructed between the two existing barriers. Alternatively, a farmer may start implementing conservation practices by constructing trash-lines across the slope to help slow runoff and trap sediment. Over time the trashlines may be replaced with grass strips, which in turn may be replaced with some form of terrace.

Understanding such incremental responses to land degradation is essential. Field professionals may only see 'half-constructed' methods of soil conservation, and conclude that the farmer had lost interest. Instead, incremental conservation is usually a deliberate response to land degradation, enabling the farmer to utilize available resources in a systematic and efficient manner, while at the same time observing whether these methods do actually work.

Table 8.2 *A Typology and Examples of Benefits of Conservation to the Land User*

Type of Benefit	Examples (two of each only)
Immediate production	Increased yield through better water conservation Less need for fertilizer because of better soils
Future production	Less risk of crop failures because of better soil quality Diversifying into higher value crops now possible on better soils
Factors of production: land, labour, capital	Increased value of the land Reduced labour needed for weeding because of better plant cover
Farming practices	Easier access onto land along terraces Stonelines a useful place to put stones and to dry weeds for composting
By-products	Poles from conservation hedgerows sold for fuelwood Grass strip vegetation cut and carried for dairy cows
Farm household	Fuelwood available from farm means less time spent walking to forest for wood Better fed cows now give milk all year, and children are healthier
Indirect, including economic and aesthetic	Greater sales of farm produce enable investment in home industry to add value to produce (eg sweet making) More wildlife attracted to farm

Bringing Together the Needed Information

Assessing the viability, technical performance and implementation of conservation technologies is probably the major task for a field professional involved in soil and water conservation. There are many hundreds of possible technologies that have the potential to reduce rates of land degradation, not least among these being indigenous technologies developed in response to local conditions. The key question is which of these technologies has the greatest likelihood of working in the biophysical environment – soil, slope, rainfall – and in the socio-economic circumstances of the land user. The most appropriate technical solution is not always suitable for the socio-economic conditions. Hence, there is the continuing

need to be assiduous in gaining a farmer-perspective.

Because the information comes from a wide variety of sources and techniques of field assessment, there is no simple quantitative way of putting it all together to obtain an overall view. Therefore, a consistent format is necessary, which highlights the important issues in summary form. Box 8.2 shows one example for *Gliricidia* contour hedgerows, which are widely employed by small farmers in South and Southeast Asia. This example describes their use in Sri Lanka. Most importantly it highlights the technology in its potential to fit the social and economic preferences of the land user.

Box 8.2 Conservation Technology Summary

Conservation Technology: *Gliricidia* single row hedges for Sri Lankan Hill Country – also called SALT (Sloping Agricultural Land Technology) throughout Asia

Description:
A live fence, acting as a barrier to sediment movement down the slope, and retaining accumulations of soil. Lines of *Gliricidia* (*G. sepium*) sticks planted along the contour of the landslope – each stick is approximately 50cm long, and planted 15–40cm apart. Each hedge is about 4–12m apart – the steeper the slope, the closer are the hedges. Dead weeds and additional sticks are sometimes placed horizontally against the planted sticks to provide a better barrier to soil movement and to stabilize the hedge.

    The sticks root quickly and within one year vigorous new growth is made. Maintenance consists of pruning new growth twice a year; placing leaves from prunings in the field to provide an organic mulch; and putting the more woody growth against the hedge to build the barrier further. Pruned sticks may also replace dead sticks. Over time, the hedges accumulate soil on the uphill side, forming bench terraces where crops are planted.

Variants/associations:
Usually drains are dug along the downhill side of the hedges. Drains may also be dug between the hedges on the terrace benches.

How does the technology work?
- A permeable barrier, trapping sediment but allowing water to pass through
- A vertical support against which up to 1.5m depth of soil accumulates
- Terraces (or less steep planting areas) form between hedges
- A source of organic mulching material from leaves which fall from prunings placed in the field
- Drains associated with hedges carry runoff

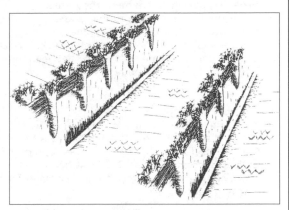

Reasons for construction and implementation:
- Conservation technology recommended for the Hill Country by local agencies and field staff
- A means of farming steep slopes with permanent barriers and fixed planting areas
- Maintenance of sufficient soil depth on slopes and long-term improvement in soil quality
- Assertion of permanent land use rights and possible means of attracting subsidies for crops
- A source of organic mulching material for fields and poles for bean supports and other uses

Other costs or opportunities forgone:
- Space taken up by hedges not available for planting
- Permanent hedges mean that most improved soil quality is close to hedges and most subject to plant competition for water and nutrients between *Gliricidia* and crops
- Upslope part of planting area may have thin soil because of downhill cultivation and digging out drainage line below hedge – hence poor crops on part of the field
- Some fertile soil buried against the hedge and is unusable by crop
- Cost of planting materials (rarely paid for, but needing labour to collect), and regular labour requirement for pruning

- Shading of crops by *Gliricidia* if not pruned regularly
- Susceptible to damage by cows.

Percentage of land taken up by technology:
Hedges only – 6 to 9 per cent; hedges plus drains – 16–25 per cent

Non-erosion benefits or opportunities gained:
- Cropping and land management: benefits gained in more accessible planting area with lower angle slope and deeper soil (on downhill part of terrace); some cultivation activities aligned along the contour parallel to hedges; soil quality improvement and greater depth enables more demanding crops; weeds can be disposed of easily by putting them in the hedges
- Economic opportunities: poles for sale or farm use; firewood for home use; subsidies available for planting and maintenance of hedges; fodder for livestock; land may be sold or leased at higher price
- Aesthetic and other benefits: hedges look impressive when intensively managed – farmer gains in reputation as a good manager

Other observations:
- Planting material usually from farm sources such as a woodlot or boundary plantings. Sticks are occasionally bought
- Labour for planting and maintenance (pruning and replacing dead sticks) is primarily a male occupation and done by the farmer
- Timely management of pruning is essential, otherwise the new growth of *Gliricidia* may easily overshadow the cropped area
- On well maintained hedges, pruned sticks are scattered in the field until the leaves drop off. The bare sticks are then woven between the living *Gliricidia* stems to form a reinforced barrier and effective sediment trap
- Hedges may offer part protection from bush pigs when combined with boundary fences

Constraints on adoption:
- The farmer does not have time to plant hedges – hedges are planted at the start of the growing season when it is raining. At this time the farmer has to work quickly to prepare the land and plant the crops. The farmer has time to plant later, but the climate is unsuitable for successful growth of the sticks
- The land is rented – where farmers rent land for only one year at a time, few will invest in planting hedges

Source: based on an unpublished paper by M Stocking and R Clarke (1997) *The Biophysical Assessment of Soil Conservation Technologies*, DFID Project R6525, Hillsides Workshop, Silsoe

Cost-benefit Analysis

Information such as appears in Box 8.2 is useful for making qualitative assessments of the possible benefits of a conservation technology to a land user, and hence its chance of acceptance. Subjective decisions on whether or not to adopt a conservation technology are likely to depend on the answer to the land user's question: 'How much money will I make (or lose) if I accept your recommendation to implement this conservation technology?' In order for the field assessor to simulate the decision-making process, monetary values must be attached to the streams of both costs and benefits and the timing at which they occur must be taken into account.

There are a number of ways of undertaking cost-benefit analysis, and a review of these is beyond the scope of this handbook. Investment appraisal – that is, the assessment of economic viability of a technology as an investment in future profitability by the farmer – is a particularly useful technique because it is relatively simple and does not demand much data beyond the sort that can be gathered by following this handbook. The benefit of conservation is then seen as if it were like an entrepreneur deciding whether to buy a new piece of equipment or take on more labour. Will the additional cost (in this case of the conservation measure) be more than made up by the additional benefit over a number of years in the economic environment that pertains locally? Appendix VI describes the steps to be followed in carrying out an investment appraisal of a conservation technology, using examples drawn from the case in Box 8.2.

The conservation technology is appraised relative to the situation that would occur if the farmer did not adopt the technology. The baseline is usually, therefore, a 'do nothing' scenario, except that the soil is allowed to deteriorate. Before the appraisal, the field data have to be assembled in a form similar to that shown in Box 8.2. Then a systematic procedure of cost-benefit analysis, followed consistently but adopting a farmer-perspective (and hence a farmer-based valuation of costs and benefits), will give the assessor a much fuller picture of:

- the primary factors in determining the magnitude of the costs and benefits from the farmer-perspective (eg cost of labour);
- the technology that stands the best chance, economically and financially, of being adopted by land users;
- the mitigation measures that might be needed (eg subsidies) for technologies that are needed for downstream (off-site) protection, which are technically efficient but economically inappropriate for the land user who is expected to implement them.

Answers to these questions are vital in the planning of any campaign for soil conservation or land rehabilitation. Unless the assessor can capture the impact of additional work in the farming system to implement conservation, then past mistakes of forcing inappropriate technologies on resentful people will just continue.

Where Do We Go From Here?

This handbook has taken field assessment of land degradation well beyond its normal confines of dedicated experts examining land for signs of deterioration in its quality and then pronouncing on the cause – usually 'poor farming' or 'improper use of land'.

Finding the causes of land degradation has been described as being like dissecting an onion. You can peel off the skin, but underneath lie successive layers that each have to be removed until the core is reached. Each layer can be seen as a cause of degradation, but at a different place, in a different scale and from a different source. So, yes, 'poor farming' may cause land degradation. But to stop at this layer implies a value-judge-ment on the part of the assessor that the farmer is to blame. Some farmers are simply bad farmers, but most are dedi-cated to their land and know well how to farm the land productively and conserva-tively. So the next layer addresses why, if the farmer is not intrinsically bad, the farming is poor. It may be that markets are insufficiently profitable to make it worth-while farming well. Or farming is simply a spare time occupation, engaged in only when other jobs are scarce. Or there is no credit to buy seeds and fertilizers … and so on. The next layer of our 'onion' should ask, 'Why is there no credit?' Eventually, the layers of the onion may reach right to causes of land degradation in the national and international economy, such as struc-tural adjustment policies imposed by the international banks, or the burden of national debt and corrupt bureaucracies. It may seem strange to many that the field assessment of land degradation may end up asking questions of geo-politics. But the reasons why land degradation occurs are extraordinarily complicated and are largely outside the control of the farmer and the field professional.

This is not to say, however, that the farmer and field professional can do noth-ing. Throughout the developing world, more and more cases are being reported of farmers who have made a success of using their land wisely and productively, despite difficult economic, social and political circumstances. The book *Sustaining the Soil* (see Appendix IV) reports 27 case studies, where farmers have developed systems of land use that are win–win – a win for themselves in providing for secure livelihoods and a win for society in keep-ing productive assets of land for future generations. All are from Africa, a conti-nent that is often seen as the most degraded and poverty ridden.

Such cases are still, sadly, not the rule. Land degradation is far too common. But the positive cases do point a way forward for more responsive, flexible and all-inclu-sive ways of dealing with land degradation, and of bringing the benefits of development to land users. If the field assessor keeps these ultimate goals clearly in mind, then he or she will be far wiser and more effective.

This handbook provides the tools for field assessors to identify the existence of, and assess the seriousness of, land degra-dation. But they go further to give guidance on how to identify the underlying causes of land degradation and to deter-mine in what way the individual circumstances of the land user are affected by decisions regarding conservation and rehabilitation. Thus, the remedies suggested are more likely to reflect the perspective of, and be more acceptable to, the land user. If the perspective of land users is thus respected, then the field asses-sor will not only have tapped into the knowledge of land users about land degra-dation, but will also have promoted more effective soil conservation and land reha-bilitation. With land users intimately involved, they will have greater ownership of the land degradation problem and the selected solution, thereby reinforcing the adoption of conservation. When field professionals work with land users, part-nerships are created which should ultimately lead to more secure futures and sustainable livelihoods.

Visual Indicators of Land Degradation

The following table summarizes the main visual indicators for the different types of land degradation. It must be remembered that these types of land degradation are interrelated. See Glossary (Appendix III) for explanations of terms.

Visual indicator	Types of soil and land degradation					
	Water erosion	Wind erosion	Salinity or alkalinity	Chemical degradation	Physical degradation	Biological degradation
Rills	✔	✗	✗	✗	✗	✗
Gullies	✔	✗	✗	✗	✗	✗
Pedestals	✔	✔	✗	✗	✗	✗
Armour layer	✔	✔	✗	✗	✗	✗
Accumulations of soil around clumps of vegetation or upslope of trees, fences or other barriers	✔	✔	✗	✗	✗	✗
Deposits of soil on gentle slopes	✔	✗	✗	✗	✗	✗
Exposed roots or parent material	✔	✔	✗	✗	✗	✗
Muddy water/mudflows during and shortly after storms	✔	✗	✗	✗	✗	✗
Sedimentation in streams and reservoirs	✔	✗	✗	✗	✗	✗
Dust storms/clouds	✗	✔	✗	✗	✗	✗
Sandy layer on soil surface	✗	✔	✗	✗	✗	✗
Parallel furrows in clay soil or ripples in sandy soil	✗	✔	✗	✗	✗	✗
Bare or barren spots	✔	✔	✔	✔	✗	✗
Efflorescence	✗	✗	✔	✗	✗	✗
Soil particles unstable in water	✗	✗	✔	✗	✗	✔
High pH	✗	✗	✔	✗	✗	✗
Low pH	✔	✗	✗	✔	✗	✔
Nutrient deficiency/toxicity symptoms evident on plants	✔	✗	✗	✔	✗	✔
Increased incidence of plant disease/morphological irregularities (eg stunting)	✗	✗	✔	✔	✔	✗
Decreasing yields	✔	✔	✔	✔	✔	✔

Visual indicator	Types of soil and land degradation					
	Water erosion	Wind erosion	Salinity or alkalinity	Chemical degradation	Physical degradation	Biological degradation
Changes in vegetation species	✔	✗	✔	✔	✗	✗
Plough pan	✗	✗	✗	✗	✔	✗
Restricted rooting depth	✔	✗	✗	✗	✔	✗
Structural degradation, including compaction	✗	✗	✔	✗	✔	✗
Poor response to fertilizers	✗	✗	✗	✔	✗	✔
Decrease in organic matter (lighter-coloured soils)	✔	✗	✔	✗	✗	✔
Increased sealing, crusting and runoff; reduced soil water	✔	✗	✔	✔	✔	✔
Decrease in number of earthworms/ants and similar	✗	✗	✗	✗	✗	✔

Appendix II

Forms for Field Measurement

Field Form: Rill

Site:
Date:

Measurement number	Width mm	Depth mm
1		
2		
3		
4		
5		
6		
7		
8		
9		
10		
11		
12		
13		
14		
15		
16		
17		
18		
19		
20		
Sum of all measurements		
Average (mm)	WIDTH =	DEPTH =

Length of rill: (m)
Contributing (catchment) area to rill: (m²)

Calculations

1   Convert the average width and depth of the rill to metres (by multiplying by 0.001).

2   Calculate the average cross-sectional area of the rill, using the formula for the appropriate cross-section: the formula for the area of a triangle (ie $^1/_2$ horizontal width x depth); semi-circle ($^1/_2\pi$x width x depth); and rectangle (width x depth). Thus, assuming a triangular cross-section, it is:

$^1/_2$ ☐ x WIDTH (m) ☐ x DEPTH (m) ☐ = CROSS-SEC AREA (m²) ☐

3   Calculate the volume of soil lost from the rill using the measured length of the rill.

CROSS-SEC AREA (m²) ☐ x LENGTH (m) ☐ = VOLUME LOST (m³) ☐

4   Convert the total volume lost to a volume per square metre of catchment.

VOLUME LOST (m³) ☐ ÷ CATCHMENT AREA (m²) ☐ = SOIL LOSS (m³/m²) ☐

5   Convert the volume per square metre to tonnes per hectare.

SOIL LOSS (m³/m²) ☐ x CONVERSION TO t/ha ☐ = SOIL LOSS (t/ha) ☐

Field Form: Gully

Site:

Date:

Measurement number	Width at lip(w_1) m	Width at base (w_2) m	Depth m
1			
2			
3			
4			
5			
6			
7			
8			
9			
10			
11			
12			
13			
14			
15			
16			
17			
18			
19			
20			
Sum of all measurements			
Average (m)	WIDTH w_1 =	WIDTH w_2 =	DEPTH (d)=

Length of gully (m)

Contributing (catchment) area to gully (m²)

Calculations

1   Calculate the average cross-sectional area of the gully, using the formula $(w_1 + w_2) \div 2 \times d$.

$^1/_2$ (AV WIDTH w_1 +AV WIDTH w_2) ☐ **x** DEPTH (m) ☐ = CROSS-SEC AREA (m²) ☐

2   Calculate the volume of soil lost from the gully using the measured length.

CROSS-SEC AREA (m²) ☐ **x** LENGTH (m) ☐ = VOLUME LOST (m³) ☐

3   Convert the volume lost to a volume per square metre of catchment.

VOLUME LOST (m³) ☐ ÷ CATCHMENT AREA (m²) ☐ = SOIL LOSS (m³/m²) ☐

4   Convert the volume lost to tonnes per hectare over the whole catchment area.

SOIL LOSS (m³/m²) ☐ **x** CONVERSION TO t/ha ☐ = SOIL LOSS (t/ha) ☐

Field Form: Pedestals

Site
Date:

Measurement number	Maximum height of pedestal in locality (mm)
1	
2	
3	
4	
5	
6	
7	
8	
9	
10	
11	
12	
13	
14	
15	
16	
17	
18	
19	
20	
Sum of all measurements	
Average (mm)	AV PED HEIGHT =

Calculations

1   Calculate t/ha equivalent of the net soil loss (represented by the average pedestal height).

AV PED HEIGHT (mm) [   ] x CONVERSION TO t/ha [   ] = SOIL LOSS (t/ha) [   ]

Field Form: Armour Layer

Site:
Date:

Measurement number	Depth of armour layer (in mm)	Proportion of coarse material in topsoil
1		
2		
3		
4		
5		
6		
7		
8		
9		
10		
11		
12		
13		
14		
15		
16		
17		
18		
19		
20		
Sum of all measurements		
Average	AL DEPTH (mm)=	COARSE % =

Calculations

1 Calculate the depth of soil required to generate the depth of the armour layer, based on the measured estimate for coarse material in the topsoil.

AL DEPTH (mm) ☐ x COARSE % ☐ = TOTAL SOIL (mm) ☐

2 Calculate the soil lost.

TOTAL SOIL (mm) ☐ – AL DEPTH (mm) ☐ = NET SOIL LOSS (mm) ☐

3 Calculate t/ha equivalent of net soil.

NET SOIL LOSS (mm) ☐ x CONVERSION TO t/ha ☐ = SOIL LOSS (t/ha) ☐

Field Form: Plant/Tree Root Exposure

Site:
Date:

Measurement number	Measured difference in soil level (A) mm	Converted to tonnes/hectare A x conversion* t/ha	Age of plant/ tree yrs	Annual change in level t/ha/yr
1				
2				
3				
4				
5				
6				
7				
8				
9				
10				
11				
12				
13				
14				
15				
16				
17				
18				
19				
20				
Sum of all measurements	–	–	–	
Average t/ha/yr	–	–	–	ANNUAL SL =

* Conversion multiplier depends on soil bulk density; see Table 4.1 or Box 4.2

Field Form: Fence Post Exposure

Site:
Date:

Measurement number	Depth of erosion (A) mm	Converted to tonnes/Hectare A x conversion* t/ha	Time elapsed since structure installed: yrs	Annual soil loss t/ha/yr
1				
2				
3				
4				
5				
6				
7				
8				
9				
10				
11				
12				
13				
14				
15				
16				
17				
18				
19				
20				
Sum of all measurements	–	–	–	
Average t/ha/yr	–	–	–	ANNUAL SL =

* Conversion multiplier depends on soil bulk density; see Table 4.1 or Box 4.2

Field Form: Waterfall Effect

Site:
Date:

Measurement number	Scoop diameter m	Scoop radius (diameter ÷ 2) r m	Scoop depth d m	Scoop volume ($^1/_3 \pi$ x r^2 x d) m^3
1				
2				
3				
4				
5				
6				
7				
8				
9				
10				
11				
12				
13				
14				
15				
16				
17				
18				
19				
20				

Total volume of soil lost from these measurements (m³)	
Average volume of soil lost from each scoop (m³)	
Field area (m²)	
Number of Scoops	

Calculations

1   Calculate the volume of soil loss for the whole field, based on the number of scoops in the field.

AVERAGE VOLUME (m³) ☐ x NO. OF SCOOPS ☐ = VOLUME OF SOIL ☐
LOST FROM FIELD (m³)

2   Calculate the soil loss per square metre, based on the measured field area.

VOLUME OF SOIL LOST ☐ ÷ AREA OF FIELD (m²) ☐ = VOLUME OF SOIL ☐
FROM FIELD (m³)   LOST PER M² (m³/m²)

3   Calculate the tonnes per hectare equivalent of this volume of soil loss.

TOTAL VOLUME OF ☐ x CONVERSION TO t/ha ☐ = SOIL LOST (t/ha) ☐
SOIL LOST (m³/m²)

Field Form: Solution Notches/Rock Exposure

Site:
Date:

Measurement number	Distance of solution notch/exposure from ground surface (mm)
1	
2	
3	
4	
5	
6	
7	
8	
9	
10	
11	
12	
13	
14	
15	
16	
17	
18	
19	
20	
Sum of all measurements	
Average (mm)	AV HEIGHT =

Calculations

1 Calculate annual change in the soil level, based on the period over which the solution notch/rock exposure occurred.

AV SOIL LOSS (mm) ☐ ÷ PERIOD OF SOIL LOSS (yr) ☐ = ANNUAL SOIL LOSS (mm/yr) ☐

2 Calculate the t/ha equivalent of the annual soil loss.

ANNUAL SOIL LOSS (mm/yr) ☐ x CONVERSION TO t/ha ☐ = SOIL LOSS (t/ha/yr) ☐

Field Form: Tree Mound

Site:
Date:

Measurement number	Measured difference in soil level (A) mm	Converted to tonnes/hectare A x conversion* t/ha	Age of plant/tree yrs	Annual soil loss due to change in surface level t/ha/yr
1				
2				
3				
4				
5				
6				
7				
8				
9				
10				
11				
12				
13				
14				
15				
16				
17				
18				
19				
20				
Sum of all measurements	–		–	
Average (t/ha/yr)	–		–	ANNUAL SL =

* Conversion multiplier depends on soil bulk density; see Table 4.1 or Box 4.2

Field Form: Build-up against Barrier

Site:
Date:

Measurement number	Measured depth m	Measured length of deposition m
1		
2		
3		
4		
5		
6		
7		
8		
9		
10		
11		
12		
13		
14		
15		
16		
17		
18		
19		
20		
Total		
Average(m)		

Length of barrier (m)
Contributing (catchment) area to barrier (m²)

Calculations

1 Calculate the average cross-sectional area of the accumulation, using the formula for the area of a triangle (ie $^1/_2$ base **x** height), where the base equates to the measured length of the deposition from the barrier into the field and the height equates to the depth of the deposition.

$^1/_2$ **x** DEPTH (m) ☐ **x** LENGTH (m) ☐ = CROSS-SEC AREA (m²) ☐

2 Calculate the volume of soil accumulated behind the barrier, based on the measured length of the barrier across the slope.

CROSS-SEC AREA (m²) ☐ **x** BARRIER (m) ☐ = VOLUME ACCUMULATED (m³) ☐

3 Convert the total volume accumulated to a volume per square metre of contributing area.

VOLUME ACCUMULATED (m³) ☐ ÷ CONTRIBUTING AREA (m²) ☐ = SOIL LOSS (m³/m²) ☐

4 Convert the volume per square metre to tonnes per hectare.

SOIL LOSS (m³/m²) ☐ **x** CONVERSION TO t/ha ☐ = SOIL LOSS (t/ha) ☐

5 Convert the total soil loss as represented by the soil accumulated behind the barrier into an annual equivalent, based on the length of time that the barrier has been in situ.

SOIL LOSS (t/ha) ☐ ÷ TIME (yr) ☐ = ANNUAL SOIL LOSS (t/ha/yr) ☐

Field Form: Build-up against Tree Trunks/Plant Stems

Site:

Date:

Measurement number	Depth d m	Distance from trunk/stem r m	Volume of soil saved $^1/_2(^1/_3 \pi \times r^2 \times d)$ m³	Contributing area m²
1				
2				
3				
4				
5				
6				
7				
8				
9				
10				
11				
12				
13				
14				
15				
16				
17				
18				
19				
20				

Total Volume Saved (m³)

Total Contributing Area (m²)

Age of Trees

Calculations

1   Calculate the annual rate of soil accumulation, based on the age of the trees/plants.

TOTAL VOLUME OF   ☐   ÷ AGE OF TREES   ☐   = ANNUAL VOLUME   ☐
SOIL SAVED (m³)   (YEARS)   OF SOIL ACCUMULATED
                                             (m³/yr)

2   Convert the total volume of soil accumulated to a volume per square metre.

ANNUAL VOLUME   ☐   ÷ CONTRIBUTING   ☐   = TOTAL VOLUME OF   ☐
OF SOIL   AREA (m²)   SOIL ACCUMULATED
ACCUMULATED (m³/yr)   (m³/m²)

3   Calculate the tonnes per hectare equivalent of this volume of soil accumulated (and, thus, the tonnes per hectare equivalent of soil lost between the tree/plant barriers).

TOTAL VOLUME OF   ☐   x CONVERSION TO t/ha   ☐   =SOIL LOST (t/ha/yr)   ☐
SOIL ACCUMULATED/
LOST (m³/m²)

Field Form: Sediment in Drain

Site:
Date:

Measurement number	Depth of sediment mm	Width of drain mm
1		
2		
3		
4		
5		
6		
7		
8		
9		
10		
11		
12		
13		
14		
15		
16		
17		
18		
19		
20		
Sum of all measurements		
Average (mm)	DEPTH =	WIDTH =

Length of drain: (m)
Contributing (catchment) area to drain: (m²)

Calculations

1 Convert the average depth and width of the sediment in the drain from millimetres to metres (by multiplying by 0.001).

2 Calculate the average cross-sectional area of the sediment in the drain.

WIDTH (m) ☐ x DEPTH (m) ☐ = CROSS-SEC AREA (m²) ☐

3 Calculate the volume of soil deposited in the drain, based on the measured length of the drain.

CROSS-SEC AREA (m²) ☐ x LENGTH (m) ☐ = VOLUME DEPOSITED (m³) ☐

4 Convert the total volume to a volume per square metre of catchment.

VOLUME DEPOSITED (m³) ☐ ÷ CONTRIBUTING AREA (m²) ☐ = SOIL LOSS (m³/m²) ☐

5 Convert the volume per square metre to tonnes per hectare.

SOIL LOSS (m³/m²) ☐ x CONVERSION TO t/ha ☐ = SOIL LOSS (t/ha) ☐

Field Form: Enrichment Ratio

Site:
Date:

Measurement number	% of fine particles in eroded soil: ie soil remaining in-field	% of fine particles in enriched soil: ie soil caught downslope and deposited
1		
2		
3		
4		
5		
6		
7		
8		
9		
10		
11		
12		
13		
14		
15		
16		
17		
18		
19		
20		
Sum		
Average	ERODED =	ENRICHED =

Calculations

1 Calculate the ratio of fine materials in the eroded soil to fine materials in the enriched soil

ENRICHED % [ ] ÷ ERODED % [ ] = ENRICHMENT RATIO [ ]

Glossary – Terms Closely Related to Assessment of Land Degradation

Aggregate stability Aggregates are groups of soil particles, also called peds. Their stability depends on the agent such as organic matter or lime which cements the particles, and on the force which breaks them down such as rainfall erosivity or the action of a plough.

Alkalinity A type of soil degradation where sodium cations increase on the exchange complex of clay and organic matter particles in the soil. Increased alkalinity leads to physical degradation as well as chemical problems.

Armour Layer The concentration at the soil surface of coarse soil particles that would ordinarily be randomly distributed throughout the topsoil.

Biomass The total weight of the organic substance and organisms in a given area. It includes growing and decaying plant materials and micro-organisms found in the soil.

Bulk Density The mass of soil divided by the volume occupied by soil, water and air.

Cation Exchange The process of ions held by electrostatic forces between the negative clay charge and positive ion charge. The most common nutrient cations are calcium, magnesium and potassium.

Cost-benefit Analysis This is a method of financial appraisal which compares the estimated future costs with the estimated future benefits of a particular course of action. The appraisal goes beyond a purely financial calculation and incorporates social advantages (such as saving time) and social disadvantages (such as noise, loss of farm land) by attaching a monetary value to them. This method of analysis seeks to mimic the subjective decision-making of individual investors, but this element of subjectivity leads to the results of such analysis being treated with some cynicism. If the net present value of the course of action is positive, then it is rational to undertake that action, whereas a negative net present value suggests that the costs of the action outweigh its benefits.

Degradation – Biological A type of soil degradation consisting of the mineralization of humus and an increase in the activity of micro-organisms responsible for organic decay, resulting in an overall decrease in organic matter.

Degradation – Chemical A number of types of soil degradation that may involve one or more of the following processes: leaching of nutritive elements; acidification; toxicities, other than excess of salts.

Degradation – Land The temporary or permanent lowering of the productive capacity of land.

Degradation – Physical A set of types of soil degradation involving one or more of the following processes: loss of soil physical structure; sealing and crusting of soil surface; reduction in permeability; compaction at depth; increase in macroporosity; limitations to rooting.

Degradation – Soil A decrease in soil quality as measured by changes in soil properties and processes, and the consequent decline in productivity in terms of immediate and future production.

Discount Rate The discount rate is an interest rate (usually a combination of current market interest rates plus an element to account for risk) applied to future cash flows to reduce (or discount) them to their current value if they were to arise immediately.

Ecosystem An ecosystem is a particular environment and the plants and animals that inhabit it. The boundaries of an ecosystem may be drawn very narrowly (eg a particle of soil) or very broadly (the entire Earth).

Efflorescence A luminous glow at night from some soil surfaces, arising from salt deposits.

Enrichment Ratio The levels of nutrients in eroded material are proportionately higher than in the source soil. The enrichment ratio measures the relative concentrations of nutrients in deposited material and in the soil from which that eroded material came. This measure does not take account of nutrients dissolved in runoff or deposited elsewhere.

Erosion The removal of soil and rock particles by the forces of water, wind, ice or gravity.

Focus Groups A focus group is a group of community residents specifically selected because they form part of a particular sub-grouping within the community, for example farmers, mothers. Discussions within the focus group should highlight local problems, knowledge, beliefs and problem-solving capacity for the sub-group.

Gully A miniature valley or gorge caused by the erosive effect of running water. The water wears away a deep channel in the land surface. Typically, water only runs through gullies after rains.

Key Informants These are community members who are particularly qualified to provide information about local conditions, usually due to their position within the community, eg local officials, community leaders, other development workers. Key informants may provide background information, or introductions to other community members or groups.

Net Present Value The value of the projected future costs and benefits of a particular course of action, discounted to the present value equivalent using an appropriate discount rate (incorporating the cost of capital and the degree of risk associated with the course of action).

Pairwise Ranking See problem ranking.

Participatory Rural Appraisal Information gathering that requires the active involvement and participation of the rural people being targeted by research and development projects. PRA involves listening to, and learning from, members of rural communities.

Pedestal A pillar of soil capped by a more resistant material (such as a stone or root) which protects the soil from rainsplash erosion.

pH The standard measure of acidity and alkalinity. High pH indicates alkalinity, often from salts. Low pH indicates acidity, often from the loss of nutrient cations.

Problem Ranking This technique allows the field worker to determine the relative importance attached to practices or problems by the community. It also helps to focus on the areas that should be prioritized by extension work. Pairs of issues are compared and a matrix constructed which records the relative importance attached to each item in each comparison. The issues can then be ordered from the most to the least important.

Resilience The ability of a land system, or a livelihood strategy, to absorb and utilize change, including resistance to a shock.

Resource Mapping This technique is a useful first step in field work as it explores the residents' perception of their community and yields information about the physical features, infrastructure, community meeting points and location of households within the community.

Rill A small channel formed on the soil surface during erosion. Rills often appear during heavy rains. They are seasonal, in that they can be eliminated by normal agricultural practices.

Salinity A type of soil degradation where salts increase in the soil water solution. It is measured by an increase in electrical conductivity.

Seasonal Calendars These record the main activities and problems which occur during the agricultural year. Cropping patterns, labour availability, food availability, water supply and health status should all be addressed in the calendar. These calendars are a good way to gain information about the farming practices in the community and an understanding of the constraints recognized by land users.

Semi-structured Interviews In this type of interview, the interviewer has a broad aim, or checklist of points to be covered, at the outset of the meeting with the interviewee. However, the answers provided by the interviewee determine the actual direction of the interview. The interviewer takes his lead from the information provided by the interviewee and explores the issue on that basis.

Sensitivity The degree to which a land system, or livelihood strategy, undergoes change due to natural forces, human intervention or a combination of both.

Shear strength The ability of a soil to resist shearing or physical breakdown. Strength is imparted to soil by cohesive forces between particles and by the frictional resistance of particles that are forced to slide over one another during, for example, tillage operations.

Social Mapping This process may be conducted as part of a resource mapping exercise, since the degree of access to resources is often determined by membership of social groups. Social linkages can be more explicitly recorded on a Venn diagram.

Soil Fertility The soil's ability to produce and reproduce. It is the aggregate status of a soil consequent on its physical, chemical and biological well-being.

Soil Productivity The overall productive status of a soil arising from all aspects of its quality and status, such as its physical and structural condition as well as its chemical content.

Subsoil The layer of soil lying immediately below the surface soil.

Time Lines This technique records changes, trends and events by reference to locally important history as remembered by the informants/community. It can help to pinpoint the causes of problems or changes.

Transect Walks These are systematic walks through the village from one boundary to the opposite boundary. Usually two walks are undertaken, perpendicular to each other (to give a cross-shaped pattern). During the walk the field worker observes the local practices, and discusses how and why things are done with the land users. Information on farming practices, access to land and water, constraints and problems should be recorded.

Topsoil The surface layer of the soil. This is the soil used for cultivation.

Wealth Ranking Information on the relative wealth (or well-being) of households in a community can be gathered where community members define how wealth (or well-being) is perceived locally, and then putting the households into order from those with the greatest level of wealth to the least. This technique is best used with individuals, but it should be carried out with at least three community members to avoid inherent biases arising due to the status of the respondents.

Well-being Ranking See wealth ranking.

Appendix IV

Annotated Bibliography

Only a sample of key publications and websites on land degradation is given here. These are the sources most useful to support the ideas and practical challenges in this handbook and to accompany training courses in field assessment techniques.

Reading on Land Degradation

Blaikie, P and Brookfield, H C (1987) *Land Degradation and Society*, Methuen, London
    – Still the standard text looking at land degradation from a non-technical perspective.
Brookfield, H (ed) (1995) 'Special Issue: United Nations University Collaborative Research Programme on People, Land Management and Environmental Change', *Global Environmental Change*, vol 5, no 4, pp263–393
    – This volume contains important PLEC papers, many of which relate to environmental change, such as land degradation. Look especially at Gyasi et al, pp355–366, on production pressures in the forest-savanna of Ghana, which provides evidence that farmers are affected by and trying to adapt to environmental change. Also, Kiome and Stocking, pp281–296, adopt a farmer-perspective to analysing the responses of farmers to erosion and conservation in semi-arid Kenya.
Fairhead, J and Leach, M (1996) *Misreading the African Landscape: Society and Ecology in a Forest–Savanna Mosaic*, Cambridge University Press, Cambridge
    – A fascinating account of the authors' investigation into the explanations for the occurrence of forest islands in the savanna transition belt in Guinea.
Lal, R, Blum, W H, Valentine, C and Stewart, B A (eds) (1997) *Methods for Assessment of Soil Degradation*, CRC Press, Boca Raton, Florida
    – This is the latest standard reference on scientific and experimental approaches to soil degradation assessment. It will appeal more to researchers than to field workers, and to those with access to soil analytical laboratories.
Oldeman, L R, Hakkeling, R T A and Sombroek, W G (1990) *World Map of the Status of Human-induced Soil Degradation* (explanatory note and world map at 1:10 million), International Soil Reference and Information Centre, Wageningen, and UNEP, Nairobi
    – This is the much-quoted GLASOD exercise, which attempts to depict soil degradation types worldwide and give global estimates of status. Remember that assessments are based almost wholly on 'expert' judgement, and certainly not on specific field indicators. Yet, this publication is highly influential in publicizing the critical status of soil and land degradation.
Scherr, S and Yadav, S (1996) *Land Degradation in the Developing World: Implications for Food, Agriculture, and the Environment to 2020*, Food, Agriculture and Environment Discussion Paper 14, International Food Policy Research Institute, Washington DC (Abstract available on-line from IFPRI at http://www.ifpri.org)
    – This was one of the first technical publications to present land degradation as a phenomenon which is not only spatially differentiated, but also an ambivalent factor in food production.
Young, A (1994) *Land Degradation in South Asia: Its Severity, Causes and Effects upon the People*, World Soil Resource Reports, United Nations Food and Agriculture Organization,

Rome (Available free on-line at http://www.fao.org/docrep/V4360E/V4360E00.htm#Contents)
– This is one of the most thorough regional assessments of land degradation, but from a
largely technical perspective.

Reading on Economic Analysis

Clark, R (1996) *Methodologies for the Economic Analysis of Soil Erosion and Conservation*,
CSERGE Working Paper, UEA, Norwich, UK
   ˙ – An excellent review of many methodologies for economic analysis in the context of land
   degradation.
Dent, D and Young, A T (1981) *Soil Survey and Land Evaluation*, George Allen & Unwin,
London
   – Chapter 11, 'The Economics of Land Evaluation', provides useful guidance on how to apply
   cost-benefit analysis, with clear tabulations of data and discounting.
Enters, T (1998) 'A Framework for the Economic Assessment of Soil Erosion and Conservation',
in F W T Penning de Vries, F Angus and J Kerr (eds) *Soil Erosion at Multiple Scales: Principles
and Methods for Assessing Causes and Impacts*, CAB International, Wallingford, UK
   – Useful paper on cost-benefit analysis, which advocates a participatory approach in order to
   best reflect the situation of the decision-maker.
Gittinger, J P (1982) *Economic Analysis of Agricultural Projects*, 2nd edn, Johns Hopkins
University Press, Baltimore
   – This is the classic text on how to undertake economic analysis, including detailed guidance
   on cost-benefit analysis of agricultural technologies.

Reading on Farmer-Perspective

Assmo, P (1999) *Livelihood Strategies and Land Degradation: Perceptions among Small-scale
Farmers in Ng'iresi Village, Tanzania*, University of Göteborg, Sweden
   – A good example of farmer perspectives.
Dejene, A, Shishira, E K, Yanda, P Z and Johnsen, F H (1997) *Land Degradation in Tanzania:
Perception from the Village*, World Bank Technical Paper No. 370, Washington DC (Cost
US$22 from World Bank Publications. Abstract on-line at
http://www.worldbank.org/html/extpb/abshtml/13993.htm)
   – A good example of determination of farmer-perspective. The following is taken from the
   Abstract:
   '*Local land users and officials often have conflicting perceptions of and responses to land
   degradation issues. This causes problems for officials in diagnosing and addressing the issue
   and is a major constraint on the successful implementation of policies and projects to address
   land degradation. This study looks at the perception and response gap between officials and
   land users in the diagnosis and remedy of land degradation. It also examines the dynamics of
   the loss of soil fertility and low productivity at the village level.*'
FAO (2000) *Guidelines and Reference Material on Integrated Soil and Nutrient Management and
Conservation for Farmer Field Schools*, Food and Agriculture Organization of the United
Nations, Rome
   – This publication describes the Farmer Field School Approach, advocated by FAO. Although
   the focus is on training farmers in farm management, farmers are central to the programme.
   They 'learn-by-doing', facilitated by extension workers, in a field setting.

Reading on Sustainable Rural Livelihoods Approach

Carney, D (ed) (1998) *Sustainable Rural Livelihoods: What Contribution Can we Make?*
  Department for International Development, London
  – This book brings together a selection of papers that outline DfID's approach to sustainable
  rural livelihoods and links this focus to the aim of the elimination of poverty.
Ellis, F (2000) *Rural Livelihoods and Diversity in Developing Countries*, Oxford University Press,
  Oxford
  – This book looks at the application of the sustainable rural livelihoods framework to devel-
  oping countries.
Scoones, I (1998) *Sustainable Rural Livelihoods: A Framework for Analysis*, Working Paper No
  72, Institute of Development Studies, Brighton, UK (Available free from the Institute of
  Development Studies on-line at http://server.ntd.co.uk/ids/bookshop/details.asp?id=419)
  – One of a stream of publications now coming out on the SRL approach and how to put it
  into practice. This approach will undoubtedly be driving the development agenda for the next
  decade; see Chapter 3 of this handbook.

Reading on Participatory Rural Approaches

Chambers, R (1994) 'The origins and practice of Participatory Rural Appraisal', *World
  Development*, vol 22, no7, pp953–969
  – The first of three articles by Robert Chambers, all published in the same volume for World
  Development (see also pp1253–1268 and 1437–1454). While the experience of PRA has now
  moved on, its basic approach remains unchanged.
FAO (2000) *Guidelines for Participatory Diagnosis of Constraints and Opportunities for Soil and
  Plant Nutrient Management*, Food and Agriculture Organization of the United Nations, Rome
  (Available free from the FAO on-line at ftp://ftp.fao.org/agl/agll/docs/misc30.pdf)
  – While claiming to be participatory, these guidelines still present a technically led approach to
  understanding natural resource management issues. Nevertheless, they offer useful checklists
  of items of information to obtain.
IIED (1995–present) *PLA Notes – Notes on Participatory Learning and Action*, International
  Institute for Environment and Development, Sustainable Agriculture Programme, London
  – Excellent series of case studies and guidance notes on conducting rural surveys.
Nabasa, J, Rutwara, G, Walker, F and Were, C (1995) *Participatory Rural Appraisal: Practical
  Experiences*, Natural Resources Institute, Chatham, UK
  – This booklet provides guidance on how to conduct PRA, with some good case studies illus-
  trating how different techniques can be employed.
Partners for Development (1999) *Field Manual for Participatory Rural Appraisal (PRA)*, Partners
  for Development, Washington DC
  – The Annex to this manual is particularly useful as it includes step by step guidance on how
  to perform different PRA techniques.

Reading on Indicators of Soil Loss

Douglas, M G, Mughogho, S K, Shaxson, T F and Evers, G (1999) *Malawi: An Investigation into
  the Presence of a Cultivation Hoe Pan under Smallholder Farming Conditions*, TCI
  Occasional Paper Series No 10, Food and Agriculture Organization of the United Nations,
  Rome
  – This booklet gives a clear account of the investigation into hoe pan in Malawi and suggests
  ways in which this problem can be tackled.

Herweg, K (1996) *Field Manual for Assessment of Current Erosion Damage*, Soil Conservation
    Research Programme, Ethiopia, and Centre for Development and Environment, University of
    Berne, Switzerland
    – Unfortunately not widely available, this little book is the closest in conception and design to
    this handbook. It takes erosion assessment in a more rigidly 'scientific' way and does not
    specifically address farmers' issues. But it is worth perusing for a more standard approach to
    field assessment of erosion processes.
Herweg, K, Steiner, K and Slaats, J (1999) *Sustainable Land Management– Guidelines for Impact
    Monitoring, Volume 1: Workbook* and *Volume 2: Toolkit*, Centre for Development and
    Environment, University of Berne, Switzerland
    – These two books describe how to monitor the impact of development projects on sustainable
    land management and provide details on a number of different tools that can be used to
    measure impact. Some of these tools are well-suited to the field assessment of land degradation
    and complement the techniques considered in this handbook.
Humphreys, G S and Macris, J L (2000) 'Some Notes on Determining Soil Loss from Exposed Tree
    Roots', *PLEC Project Report* (unpublished notes available from the first-named author)
    – These notes provide useful lists of tropical and sub-tropical trees that display annual and
    twice-annual tree rings.
Lal, R (ed) (1994) *Soil Erosion Research Methods*, 2nd edn, Soil and Water Conservation Society,
    Ankeny, Iowa
    – A useful book of rather standard approaches to measurement and assessment. The Cover
    Model mentioned in Chapter 7 of this handbook is dealt with on pp210–232 under 'Assessing
    Vegetative Cover and Management Effects'.
Landon, J R (ed) (1984) *Booker Tropical Soil Manual: A Handbook for Soil Survey and
    Agricultural Land Evaluation in the Tropics and Subtropics*, Booker Agriculture International
    Limited, London
    – A valuable reference book giving information on soil characteristics and classification. It also
    gives guidelines on the suitability of soils for major tropical crops.
Stocking, M A and Clark, R (1999) 'Soil Productivity and Erosion: Biophysical and Farmer-
    perspective Assessment for Hillslopes', *Mountain Research and Development*, vol 19, no 3,
    pp191–202
    – This paper describes a number of field techniques for erosion rate and impact assessment.
    Some examples in this handbook are drawn from this source. It contains a comprehensive
    reference list.

Reading on Soil Loss Estimation Models

Elwell, H A and Stocking, M (1982) 'Developing a Simple yet Practical Method of Soil Loss
    Estimation', *Tropical Agriculture*, vol 59, pp43–48
    – This paper describes the SLEMSA model, and shows a worked application for smallholder
    farming in Zimbabwe.
Morgan, R P C (1995) *Soil Erosion & Conservation*, 2nd edn, Longman Group, London
    – See Chapter 4 for discussion of erosion hazard assessment.
Renard, K G, Foster, G R, Weesies, G A, McCool, D K and Yoder, D C (1997) 'Predicting Soil
    Erosion by Water: A Guide to Conservation Planning with the Revised Universal Soil Loss
    Equation (RUSLE)', *Agricultural Handbook 703*, Agricultural Research Service, United States
    Department of Agriculture, Washington, DC
    – This publication provides guidance on the use of the Revised Universal Soil Loss Equation
    and, as such, is an update on the classic text (Wischmeier and Smith, noted below).
Stocking, M and Peake, L (1986) 'Crop Yield Losses from the Erosion of Alfisols', *Tropical
    Agriculture*, vol 63, no 1, pp41–45

– This paper brings together the evidence for yield decline with erosion, illustrating it by a simple model of soil productivity.

Wischmeier, W H and Smith, D D (1965) 'Predicting Rainfall Erosion Losses from Cropland East of the Rocky Mountains', *Agricultural Handbook 282*, Agricultural Research Service, United States Department of Agriculture, Washington, DC

– This is the classic booklet on the design, operation and statistical basis of the Universal Soil Loss Equation.

Reading on Land Quality Indicators and Nutrient Deficiencies

Havlin, J L (1999) *Soil Fertility and Fertilizers: An Introduction to Nutrient Management*, Prentice-Hall, London

– This book provides clear information on nutrient deficiencies and other factors that affect soil quality.

Hilhorst, T and Muchena, F (eds) (2000) *Nutrients on the Move: Soil Fertility Dynamics in African Farming Systems*, Drylands Programme, International Institute for Environment and Development, London

– This little book makes the case for a gradual decline in nutrient and organic matter status of African farming systems and then shows from a farmer perspective how farmers are coping and, sometimes, reversing the situation. Excellent insights into farmers' practices.

UN/FAO (1997) *Land Quality Indicators and Their Use in Sustainable Agriculture and Rural Development*, FAO Land and Water Bulletin 5, co-published with World Bank, UNEP and UNDP, Rome

– A thorough review of recent efforts to develop indicators of land quality. Unfortunately, most authors in this FAO paper adopt a standard 'scientific' approach – but it makes a good contrast to farmer-first perspectives.

Reading on Plant Analysis

Wallace, T (1961) *The Diagnosis of Mineral Deficiencies in Plants by Visual Symptoms: A Colour Atlas and Guide*, HMSO, London

– Although a little old, this publication provides useful guidance on how to test for nutrient deficiencies by selecting the appropriate acid/reagent.

Reading on Consequences of Land Degradation for Land Users

Douglas, M (1997) *Guidelines for the Monitoring and Evaluation of Better Land Husbandry*, Association for Better Land Husbandry (Contact T F Shaxson, Chairman, ABLH, Greensbridge, Sackville Street, Winterbourne Kingston, Dorset DT11 9BJ, UK)

– This booklet proposes and gives examples of grading mechanisms for use in evaluating the seriousness of land degradation.

Nabhan, H, Mashali, A M and Mermut, A R (eds) (1999) *Integrated Soil Management for Sustainable Agriculture and Food Security in Southern and East Africa*, Land and Water Development Division Publication AGL/MISC/23/99, United Nations Food and Agriculture Organization, Rome (available from FAO Publications)

– See especially pp91–120, by Stocking and Tengberg, where erosion-induced loss in soil productivity is shown to have an impact on agricultural production and food security. Land degradation outcomes in Chapter 7 come from this publication.

Tengberg, A, da Veiga, M, Dechen, S C F and Stocking, M (1998) 'Modelling the Impact of Erosion on Soil Productivity: a Comparative Evaluation of Approaches on Data from Southern Brazil', *Experimental Agriculture*, vol 34, pp55–71

– This paper demonstrates a number of techniques and approaches of presenting erosion-productivity data. The modelling described is intended to provide planners and policy-makers with predictions for the sustainability of soil resources.

Reading on Potential Benefits of Conservation

Hurni, H (1986) *Guidelines for Development Agents on Soil Conservation in Ethiopia*, Community Forests and Soil Conservation Development Department, Ministry of Agriculture, Ethiopia
– This small book deals specifically with conditions in Ethiopia, but it gives useful summaries of several conservation measures and describes the types of landscape to which they are most suited.

Reij, C, Scoones, I and Toulmin, C (1996) *Sustaining the Soil: Indigenous Soil and Water Conservation in Africa*, Earthscan, London
– An excellent collection of case studies, demonstrating farmers' rationale in undertaking soil and water conservation.

Websites

http://www.cde.unibe.ch/programmes/global/glo20.html
This is WOCAT – the World Overview of Conservation Approaches and Technologies, which includes a comprehensive database of conservation techniques from many developing countries. A CD-ROM is also available from the project's coordinators at the University of Berne. It is relevant for this handbook in that WOCAT shows the factors and variables important to farmers in determining whether or not to invest in land rehabilitation.

http://www.fao.org/WAICENT/FAOINFO/AGRICULT/agl/agll/oldocsl.asp
A useful page on the FAO website giving on-line documents on 'Land', including several of the older FAO Soils Bulletins, such as *Framework for Land Evaluation*, which could be useful in making assessments of the potential for land degradation.

http://www.ids.ac.uk/blds
The British Library for Development Studies, based at the Institute for Development Studies, Sussex. The library catalogue allows free access to many useful publications, especially in the field of Sustainable Rural Livelihoods.

http://www.unccd.int/main.php
This is the main website of the United Nations Secretariat of the Convention to Combat Desertification (CCD). The CCD has the mandate for 'desertification', which they define as land degradation in dryland areas. It is a new site and is expanding to include lists of related websites, including national inventories of desertification.

http://www.unu.edu/env/plec
This is the project website of People, Land Management and Environmental Change (PLEC). PLEC works on demonstration sites in 12 tropical countries, where local people manage their land in such a way as to preserve a large amount of biodiversity. They do this partly by managing land in sensitive ways so that it does not become degraded. The website contains a number of downloadable publications from PLEC that address field assessment of agricultural biodiversity. PLEC includes land rehabilitation and conservation assessment under the overall heading of 'agrodiversity'.

Major Tropical Soils and their Susceptibility to Land Degradation

Only those soils that are mentioned in this publication by their FAO classification (1974 Soil Map of the World edition) name and/or those that are widespread in tropical environments are described.

FAO-UNESCO soil name: soil unit and subunit	US Soil Taxonomy name	Main properties and susceptibility to land degradation

Acrisols
• orthic
• ferric
• humic
• plinthic
	Ultisol
	[orthic=Hapludults;
	ferric=Palexerults;
	humic=Humults;
	plinthic=
	plinthudults]
		Acid, low base status (<50 per cent base saturation) and strongly leached. One of the most inherently infertile soils of the tropics, becoming degraded chemically and organically very quickly when utilized. An orthic Acrisol in Indonesia on a 13 per cent slope under 3000mm rainfall has been recorded as having over 260t/ha/yr erosion. All nutrients, except Al, decreased substantially. Acrisols have very low resilience to degradation and moderate sensitivity to yield decline

Andosols • ochric • mollic • humic • vitric	Inceptisols – Andepts [ochric & humic= Dystrandepts; mollic=Eutrandepts]	From volcanic ash parent material; high in organic matter. Highly erodible, and limited in phosphorus. Chemical fertility is variable, depending on degree of weathering. Andosols have low resilience, and variable sensitivity
Arenosols • cambic • luvic • ferralic • albic	Entisols – Psamments	Consists of unconsolidated wind-blown or water-deposited sands. One of the most inherently infertile soils of the tropics and subtropics with very low reserves of nutrients. Yet if chemical inputs provided, they yield well. Arenosols have moderate resilience and low sensitivity
Cambisols • eutric • dystric • humic • calcic • chromic • vertic • ferralic	Inceptisols [eutric and calcic= Eutrochrepts; dystric= Dystrochrept; humic= Haplumbrepts; vertic=Vertic Topepts; ferralic=Oxic Tropepts]	Tropical 'brown earth' with a higher base status than Luvisols, but otherwise similar limitations. They have relatively good structure and chemical properties, and are not therefore greatly affected by degradation processes until these become large. Because of increasing clay with depth, they tend not to be greatly impacted by degradation. Cambisols have high resilience to degradation, and moderate sensitivity to yield decline

Ferralsols
• orthic
• xanthic
• rhodic
• humic
• acric
• plinthic
	Oxisols
	[orthic, xanthic
	and rhodic =Orthox;
	humic=Humox;
	acric=Acrox;
	plinthic=
	Plinthaquox]
		Ferralsols are the classic red soils of the tropics, because of high iron. They have low supply of plant nutrients and are not therefore impacted greatly by erosion; they have strong acidity and low levels of available phosphorus. With very few reserves of available minerals and easily lost topsoil organic matter, Ferralsols have low resilience and moderate sensitivity

Fluvisols • eutric • calcaric • dystric • thionic	Inceptisols – Fluvents [thionic= Sulphaquept or acid sulphate]	Formed from unconsolidated water-borne materials. Highly variable, but much prized for intensive agriculture. Under most conditions they have high resilience and low sensitivity. The big tropical exception is acid sulphate soils, which have massive chemical degradation impacts when drained for agriculture
Histosols • eutric • dystric	Histosols	Organic or peat soils. When drained, highly prized for agriculture. Land degradation often caused through shrinkage of the organic matter and subsidence
Luvisols • orthic • chromic • calcic • vertic • ferric • plinthic	Alfisols [orthic=Hapludalfs; chromic= Rhodexeralf; calcic=Haplustalf; vertic=Vertic Haploxeralfs]	The tropical soil most used by small farmers because of its ease of cultivation and no great impediments. Base saturation >50 per cent. But they are greatly affected by water erosion and loss in fertility. Nutrients are concentrated in topsoil and they have low levels of organic matter. Luvisols have moderate resilience to degradation and moderate to low sensitivity to yield decline
Nitosols • eutric • dystric • humic	Alfisols and Ultisols [eutric=Tropudalfs; dystric=Tropudults; humic=Trophumults]	One of the best and most fertile soils of tropics. They can suffer acidity and P-fixation, and when organic carbon decreases, they become very erodible. But erosion has only slight effect on crops. Nitosols have moderate resilience and moderate to low sensitivity
Phaeozems • haplic • calcaric • luvic	Mollisols [haplic=Hapludolls; calcaric=Vermudolls; luvic=Argiudolls]	They have a good structure and are generally resistant to erosion. But once eroded, the effect on yields is great. They have a high resilience and high sensitivity
Rendzinas	Mollisols – Rendolls	Characterized by extreme shallowness, and formed on limestone (calcareous) parent material. Degradation serious with severe limitations imposed by depth and high permeability
Solonchaks • orthic • mollic	Aridisol – Salorthid	Soils having high content of salts, common in arid and semi-arid areas. Badly run irrigation schemes may turn soils into solonchaks
Solonetz • orthic	Alfisol – Natrustalf	Soils have severe chemical problems associated with salt and sodium on the exchange complex. They degrade very easily, and large gullies typically form
Vertisols • pellic • chromic	Vertisols [pellic=Pelluderts; chromic= Chromudert]	Soils with 30 per cent or more clay. Clays usually active, cracking when dry and swelling when wet. Extremely difficult to manage (hence easily degraded) but very high natural chemical fertility if physical problems overcome
Xerosols	Aridisols	Soils of the deserts, with low levels of organic matter. Subject to wind erosion and concentration of soluble salts
Yermosols	Aridisols	Even drier and more problematic than Xerosols

Investment Appraisal

Adapted from an unpublished paper by R Clark: 'Investment Appraisal of Soil Conservation Technologies: Guidance Notes and A Worked Example', November 1997.

The following ten steps give a suggested approach to investment appraisal. They are given only in outline form to illustrate the sequence; for more information, the reader is referred to any standard text on cost-benefit analysis.

Step 1: Define the 'with' and 'without' technology situations

A systematic description is needed of the technology to be appraised. How does it function? What does it do? What materials are needed to implement it? And so on.

In the example in Box 8.2 (Chapter 8), the 'with technology' situation is single-row Gliricidia *hedgerows planted across the contour. The 'without technology' situation is steep-slope arable cropping without any direct measure of keeping soil on the slope. The materials for the technology are as described in the box.*

Step 2: Convert the data into common units

Usually it is sensible to convert field areas into hectares, and yields into kilograms per hectare, although locally relevant measures may also be used. Money should be in local currency terms, with values reflecting real values and real costs to the land user. So, crop revenues should be calculated based on the price paid to farmers for their crops – the producer price – not the price at which they can be bought in the market – the market price. Inflation is a major problem in many countries, so a fixed date for valuation will usually need to be specified.

Farmers in Sri Lanka do not use field measurements of hectares and they measure their crops in the number of sacks taken from the field. Traders come to their farms to buy produce – so the crop revenues are valued according to the prices paid by traders.

Step 3: List the costs and benefits

This is the first vital step in bringing the information into some common format – two columns representing costs to the land user and benefits. Field observations and data collected from farmers are vital in undertaking this listing. The list should include only costs and benefits that occur as a result of adopting the technology. Any cost or benefit that would also occur if the farmer did not adopt the technology should not be included. Double-counting of benefits must be avoided.

Costs and benefits are derived from checklists developed from this handbook, but adapted for the specific circumstances of hill land farmers. The by-products of conservation technologies are especially important, such as the leaves of Gliricidia *hedgerows used as a mulch on the fields. Labour constraints are also a major consideration, since farmers have to balance their steep slope farming with their rice paddies. If the season is difficult, they may even abandon steep slope farming altogether. To avoid double-counting, the impact of erosion on yields and the value of the increased soil quality cannot both be taken – they represent different facets of the same process of improvement.*

Step 4: List the monetary values for each cost and benefit

The monetary values must be based on the costs and benefits to the land user, expressed usually in local currency (such as rupees) per hectare. Costs and benefits for which there are no monetary values are usually excluded.

The production benefits and the costs due to forgone opportunities are the principal monetary values, which can be relatively easily priced by the farmer. For example, Gliricidia *hedges take up land space that cannot then be used for cropping. This cost (or production opportunity forgone) may be valued as the value of the crop less a percentage that reflects the proportion of land taken up by the hedgerow.*

Step 5: Identify the ranges in data to be used in the appraisal

One of the commonest mistakes is to assume that rural society is homogeneous and that all farmers have the same perspectives. Different farmers have different values and they give responses accordingly. This variation needs to be reflected in terms of minima and maxima – ie ranges in value that encompass the spread. These ranges are then used for further calculation, and they will identify especially where some farmers may gain a net benefit and others a net cost because of their different circumstances.

In the Sri Lankan hills, yields of vegetables can be very different; some farmers grow very sizeable surpluses for sale, while others have barely enough on which to subsist. Yields were found to vary according to status of land degradation, but also according to management, soil type and responsiveness of farm practices to crop needs (eg supplementary irrigation). These factors determine a large range in value of produce, and only the more productive farmers may get a net benefit through implementing the conservation technology. In other words, the investment is only worthwhile for good farmers on good soils.

Step 6: Identify the time period for the appraisal

The time period may be the life of the technology itself, as recognized by farmers, or it may be the number of years over which farmers assess it as an investment in improving their land. Time period has important implications because improvement in land quality happens slowly, so some benefits may only be realized after the life of the technology.

The Gliricidia *hedgerows are typically replaced every 7 to 20 years, depending on the vigour and growth of the trees, and the accumulation of soil. Commonly, farmers uproot their hedgerows and plant new lines on the old terraces. They reason that the soil accumulated behind the old* Gliricidia *is fertile and should be used productively. The time-period of the appraisal in this instance should be the life of the hedge and could safely be taken as the upper limit of 20 years.*

Step 7: Construct a summary table

The table should have years listed in the first column, with a row assigned to each year of the appraisal. The body of the table is then devoted to two main sections for costs and benefits, with two columns for each type of cost or benefit to accommodate the range of values from the minimum to the maximum. If actual and relatively unchangeable costs are known for some items, then these are used.

An example of a summary table for Gliricidia *hedgerows is given here. Costs and benefits are specified in local currency at prevailing prices to the farmer. So fertilizer 'benefit' is priced at the price delivered at the farm gate. The values a to k will be used in the next step.*

Year	COSTS (and resources required)					BENEFITS					
	Labour		Tools	Loss in crop area		Increase in crop yield		Saving on fertilizer		Pole production	
	Min	Max	Actual	Min	Max	Min	Max	Min	Max	Min	Max
1											
2	a	b	c	d	e	f	g	h	i	j	k
3											
etc											

Step 8: Calculate total costs and benefits, and net cash flow for each year

The minimum and maximum data are kept separately. So for both total cost and total benefit, a minimum and maximum value is calculated for each year. The net cash flow is then calculated for each year by subtracting total costs from total benefits.

From the summary table for Gliricidia *hedgerows, total costs, benefits and net cash flow are entered. The items a to k at Step 7 show how the data are ordered. Note especially that minimum net cash flow equals minimum total benefits minus maximum total costs. Similarly, maximum net cash flow equals maximum total benefits minus minimum total costs.*

Year	TOTAL COSTS		TOTAL BENEFITS		NET CASH FLOW	
	Min	*Max*	*Min*	*Max*	*Min*	*Max*
1						
2	$a+c+d=r$	$b+c+e=s$	$f+h+j=t$	$g+i+k=u$	$t-s$	$u-r$
3						
etc						

Step 9: Adjust the net cash flow for the time value of money

The time-value of money is involved in investment appraisal because sums of money are received (benefits) and spent (costs) at different points in time. The sums of money are multiplied by a factor that is related to 'discount rate', which expresses how the value of money diminishes over time. The appraisal reflects only the value now, or 'net present value'. So a benefit in the future is worth less than a benefit now. A cost in the future is worth less at the present time than a cost now. Because discount rates are often difficult to fix and depend on external factors such as the cost of borrowing money, it is good practice to set a lower and upper discount rate and to use both of these in the calculations; see final step 10.

> Gliricidia *hedgerows and their associated terraces demand a lot of labour to plant and to construct initially. Then there are some maintenance costs in pruning the hedges and replanting trees that have died, but this is relatively small in cost. Benefits, however, come only slowly. The soil improves in quality only after a long time, having to recover from the initial earth movement in making the terraces. So, with the costs coming early and the benefits coming late, the adjustment for net cash flow for the time value of money means that very few farmers will find investing in these hedgerows financially worthwhile. Maybe only farmers who are retired employees with other sources of income can afford them.*

Step 10: Calculate the net present value of the technology

The net present value (NPV) is calculated by adding the present values of the net cash flow for each year of the appraisal. The upper and lower discount rates and the minimum and maximum discounted cash flows should be kept separate. The discount factor is derived from standard tables – the further into the future, the smaller is the factor to account for the lower net present value of money as time progresses. NPV then is the sum of discounted net cash flows over the period of the appraisal. If NPV is positive, it indicates that at that discount rate the benefits of the investment exceed the costs. So the investment is economically worthwhile at that discount rate. Alternatively, if NPV is negative, the investment is not economically viable. Conservation technologies with negative NPV are very unlikely to be acceptable to land users because, to implement them, the land user would be poorer. Because the whole appraisal has been carried out with ranges of data (minimum/maximum; upper/lower discount rate) there will be several answers, ranging from a best to a worst case scenario.

> *The final table brings all the calculations together. Note the several different values for NPV, ranging from best case scenario (maximum discounted net cash flow at the lower discount rate) to worst case scenario (minimum discounted net cash flow at upper discount rate)*

Year	Lower discount rate			Upper discount rate		
	Discount factor	*Minimum discounted net cash flow*	*Maximum discounted net cash flow*	*Discount factor*	*Minimum discounted net cash flow*	*Maximum discounted net cash flow*
1						
2						
3						
etc						
NPV Total	–			–		

Suggested Outline for a Two Week Training Workshop in Land Degradation Field Assessment

This suggested outline is based on an actual workshop conducted in Uganda with 20 participants and two facilitators, used to test and evaluate the techniques in this handbook. One of the major outcomes was that ten (rather than five) days were needed to gain a full appreciation of the techniques and their application to designing conservation approaches with land users.

Day	Activities
Day 1: Introduction to Land Degradation, Sustainable Rural Livelihoods and PRA Tools	Welcome and introduction of participants and facilitators
	Identification of workshop objectives
	Discussion of land degradation – not to be looked at in isolation but as an obstacle to sustainable livelihoods – direct impact on people
	Overview of field assessment techniques and their potential usefulness to practitioners
	Split of workshop participants into working groups
	Introduction to Sustainable Rural Livelihoods framework
	Differentiation of the capital components in sustainable rural livelihoods, with particular reference to the expected circumstances in the field site selected for the workshop activities
	Discussion on collection of data in the field – both physical measurements and information from local people
	Overview of PRA tools and how they help in the collection of information relevant to land degradation
	Role play (by workshop participants) to demonstrate both good practice and some of the pitfalls to be avoided in using PRA tools
	Group work: planning of field visit – identification of the types of information required and the tools that may be useful to obtain these data
	Group presentations to workshop of their work programme for Day 2
Day 2: Site Investigation and Familiarization	Field visit
	On-site meetings with farmers to determine perceived problems
	Site familiarization and transect walks with farmers
	Group work – preparation of a site map and summary of the capital assets of the farmer
	Introduction to cost-benefit analysis – how this can be used as a tool to compare the 'do nothing' scenario with different conservation (or alternative use) options
Day 3: The SRL Framework Applied and Demonstration of Techniques in Chapter 4	Group presentations to workshop on site investigation and SRL framework as applied to farmer
	Discussion of risks identified in each of the farms
	Identification of omissions from site maps or farmer discussions
	Training site visit – demonstration of some of the techniques described in Chapter 4 on soil loss indicators
	Introduction to triangulation

Day 4:
Measurement
of Soil Loss

Field visit
Group work to measure soil loss using the field assessment techniques set out
in Chapter 4. Data should be recorded onto Field Forms
Group work – calculation of soil loss using field assessment methods
Group presentations of estimated soil loss on individual farms and discussion
of differences in findings (if any) between methods

Day 5:
Production
Constraints

Training site visit
Introduction to methods of identifying production constraints
Field visit
Group work to assess the existence and seriousness of production constraints
Group work – evaluation of identified production constraints
Group presentations of production constraints on individual farms and
discussion of differences in findings (if any) between methods

Day 6:
Combining
Indicators of
Soil Loss,
Production
Constraints and
SRLs

Further discussion of triangulation – in the context of the different types of
data collected
Group work to combine the various data collected on soil loss and production
constraints to determine trends in land degradation on individual farms
Group work to assess how capital assets within the SRL framework contribute
to the picture of degradation or conservation in light of the physical evidence
on the farm
Group presentations on the consistency (or otherwise) of the combined
indicators

Day 7:
Semi-
quantitative
Assessments
and Erosion/
Time/Yield
Relationships

Discussion of scoring methods as ways of quantifying the effects of land
degradation, and consequently its seriousness
Group work to develop scoring mechanisms to rank the seriousness of land
degradation (or the effectiveness of conservation) on the different field units
within individual farms
Discussion between groups to determine whether scoring methods developed
are transferable or applicable only to the related farm
Introduction to erosion/time/yield relationships
Group work to determine whether, based on the data that have been collected,
any conclusions can be drawn on the relationships between erosion, yield and
time on the individual farms

Day 8:
Cost-Benefit
Analysis and
Conservation
options

Further discussion on cost-benefit analysis and how this can be used to
determine responses to conservation interventions
Discussion of conservation and rehabilitation options, and investigation of the
costs and benefits associated with these kinds of options
Group work to determine the most appropriate conservation or rehabilitation
interventions for the individual farmer and farm

Day 9:
Investment
Appraisal

Explanation of the steps in investment appraisal
Group work to convert the findings about individual farmers and their farms
into a comprehensive investment appraisal
Group work to prepare a final report on the likelihood of adoption of
conservation or rehabilitation measures by the individual farmer

Day 10:
Conclusion

Presentation of group reports (approximately one hour per group, summariz-
ing the background information gathered during the workshop and how this
combines to give a balanced picture of the likelihood of the farmer to
conserve or degrade and making final recommendations which it is considered
are appropriate and achievable for the farmer)
General discussion of method from initial site visit to final recommendations
Workshop evaluation

Notes:

1 The smaller the number of participants in the workshop, the greater the opportunities for the participants to learn the techniques suggested in the handbook. An absolute maximum number of participants should be 20, with two facilitators. If only one facilitator is available, then the participant numbers should be reduced.

2 When splitting the workshop into working groups, try to get a reasonable balance of age, seniority, experience and men and women in each group. Where a number of different languages or dialects may be spoken in a region, particular attention should also be given to ensuring that at least one group member has the ability to communicate directly with the farmer.

3 This workshop requires the willing participation of a number of farmers – one farmer (and farm) for each working group. The facilitators should visit these farms and farmers in advance of the workshop to ensure their suitability. In particular, evidence of measurable soil loss or movement is important. It is preferable to have standing crops in the field (perennial crops will do), so that the workshop participants can consider crop growth characteristics and nutrient deficiencies in addition to crop yield. Payment should be offered to the farmer to compensate for the time taken, as the workshop participants will spend at least three days in the field. Provision should also be made to compensate the farmer in the event that crops are damaged during the field training.

4 Consideration needs to be given to the distance between the workshop venue and the selected farms. If the farms are too close to a town, agriculture may be less important than off-farm employment, thus skewing the data collected by the participants. However, if the farms are too far from the workshop venue, a lot of time will be wasted travelling to and from the field site.

5 It may be advisable to identify an additional site (possibly closer to the workshop venue) which can be used for the training/demonstration sessions on Days 3 and 5.

6 It is assumed that workshop participants are already familiar with Participatory Rural Appraisal and Participant Learning and Action. Thus, only a short time is allocated to give a brief overview of the tools that may be used in the field. If the participants are not familiar with these techniques, then additional time will need to be set aside to go through them in much greater detail.

7 The workshop is designed to take a full 10 days. No sessions are scheduled for the evenings, but it is likely that as the groups start to work together on the various elements of the course that they will want to continue working into the evening. Therefore it is suggested that all the workshop participants stay as close as possible to the workshop venue.

8 Provision should be made on Days 2, 4 and 5 for the course participants to take their lunch in the field, as the investigations on each of these days can take many hours.

9 Regular feedback sessions should be held to determine how the participants are finding the workshop content and timetable.

Notes

Chapter 3

1 The Sustainable Rural Livelihoods approach has been developed by the UK's Department for International Development, particularly for use in natural resources projects. The framework has been designed for the analysis of livelihoods. It aims to incorporate the many and varied strands of rural livelihoods and to recognize the interactions and changes between these strands

2 We are grateful to Christine Okali for allowing us to base this part on her unpublished teaching notes

Chapter 4

1 This effect can be allowed for when the volume of the tree mound (V_M) exceeds the volume of the lifted soil which can be assumed to be the volume of the root bole (V_B), which is approximately given as the product of the basal tree area and its diameter. When $V_B \geq V_M$ the original soil height (H_0) is assumed to be the height at the edge of the existing mound. Otherwise when $V_M > V_B$, $H_0 = (V_M - V_B)/0.33 \, \pi \, r_M^2$ where $V_M = 0.33 \, \pi \, r_M^2 \, h$ and $V_B = \pi \, r_M^2 \, d$, and where h is the height of the tree mound and d is the basal diameter of the tree

Chapter 7

1 Further analysis of these matters is outside the scope of this handbook. The reader is referred to texts in the broad field of political economy of natural resources, such as Blaikie and Brookfield's *Land Degradation and Society* (see Annotated Bibliography – Appendix IV)

2 This sensitivity–resilience matrix is further developed in Chapter 8 in order to express how the outcomes for land users can be developed into conservation interventions. This is based on the understanding of how easy it is to degrade the land and how easy it is to restore – see Table 8.1

Index

Page numbers in *italics* refer to forms for field measurement